Mastering The OTDR
Trace Acquisition And Interpretation

Version 1.0

Eric R. Pearson, CFOS

Pearson Technologies Inc.

Acworth, GA

Disclaimer

All instructions contained herein are believed to produce the proper results when followed exactly with appropriate equipment. However, these instructions are not guaranteed for all situations.

Notice To The Reader

Trademark Notice

All trademarks are the property of the trademark holder. Trademarks used in the installation documents include Kevlar™ (DuPont), Hytrel™ (DuPont), and ST-compatible (Lucent).

Published by

Pearson Technologies Inc.
4671 Hickory Bend Drive
Acworth, GA 30102
770-490-9991
www.ptnowire.com
fiberguru@ptnowire.com

File: MTO-v1.1

Printed in the United States of America.

10 9 8 7 6 5 4 3 2 1

9781466429291

TABLE OF CONTENTS

AUTHOR'S PREFACE

TABLE OF FIGURES

TABLE OF TABLES

AUTHOR'S PREFACE

For 34 years, I've been working in fiber optic communications. I've made or viewed more than twenty one thousand OTDR races. During these experiences, I, and several of my professional associates, have noted that OTDR testing and interpretation are the two aspects of installation that cause the most difficulty to the largest number of novice installers. This book is designed to reduce such difficulty significantly! With diligence, this book helps you eliminate this difficulty completely!

My strategy is 'divide and conquer'. This book divides the essential knowledge and understanding into five, clearly written, concise, yet comprehensive chapters. Since words alone will not be sufficient, each chapter includes figures, 120 in all, to ensure that all concepts are clear. To further assist you, each of these five chapters includes a summary of key concepts, a total of 66.

These five chapters guide you through development of the understanding you need to make and interpret OTDR traces properly. The sixth chapter presents a brief summary of steps you take during field-testing. This summary includes the easily overlooked practical aspects of OTDR testing. The 151 review questions and exercises of the seventh chapter further assist you in testing, developing, verifying and strengthening your understanding. The appendices contain answers or locations of answers.

But this text includes more than what I know you need to understand. The 8421 people I've trained in more than 500 presentations have asked many excellent questions. These questions have enabled me to refine my explanations so that almost anyone can understand the concepts. Finally, these questions have defined the content and structure of this book.

The goal of this book is to develop your knowledge, abilities, and confidence to make and interpret traces properly. With this confidence, you will, rightly so, consider yourself a

Master Of The OTDR!

Best Regards,

Eric R. Pearson, CFOS/T/C/S/I

Pearson Technologies Inc.

fiberguru@ptnowire.com

770-490-9991

http://www.ptnowire.com/

October 2011

1   OTDR PRINCIPLES

Chapter Objective: you learn the basic concepts and principles by which an OTDR creates a trace.

1.1   BASIC CONCEPTS

The OTDR operator uses an OTDR to test the loss along an optical fiber link. A link is the transmission path between a transmitter and receiver. It consists of one or more segments. Each segment can consist of cable, splices, and connectors (Figure 1-1).

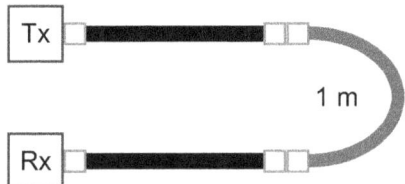

Figure 1-1: Simple Link

1.1.1   INTRODUCTION

The OTDR has two functions:

➢ To launch a pulse of light into the optical cable

➢ To measure optical power that returns to the OTDR from the cable under test

It measures that returning power as a function of the time it takes an input pulse to travel to a location and return to the OTDR. The OTDR displays the returning power level as a function of fiber distance by calculating this distance from the round trip travel time.

To use an OTDR properly, the operator needs understanding of eight concepts:

➢ Fiber structure

➢ Wavelength

➢ Attenuation

➢ Backscatter

➢ Backscatter coefficient

➢ Index of refraction

➢ Fiber length vs. cable length

➢ Reflections

1.1.2 FIBER STRUCTURE

An optical fiber structure has three regions:

➢ Core

➢ Cladding

➢ Primary coating

Most of the optical energy travels in the core, the center of the fiber. Precise alignment of cores at patch panels and splices results in low loss connections.

The core diameter, in microns (μ), can be large or small. In a data communication fiber, a large core has a diameter of either 50 or 62.5 μ (Figure 1-2). Such a fiber is a graded index (GI) multimode fiber. In data communication, telephone, and CATV networks, a small core has a diameter of 8-10 μ. Such a fiber is a singlemode fiber (Figure 1-3).

The operator uses a launch cable (4.1.2) with the same core diameter as that in the cable under test. Such matching results in low loss at the connection between the two cables.

The cladding has two functions:

➢ Confinement

➢ Increase in size

A= core= 50 or 62.5 µm

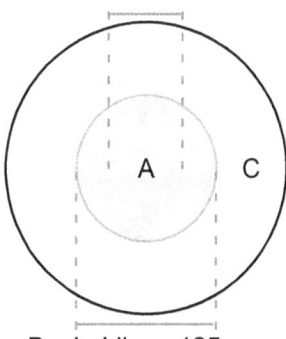

B=cladding= 125 µm
C= primary coating

Figure 1-2: Multimode Fiber

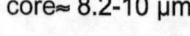

core≈ 8.2-10 µm

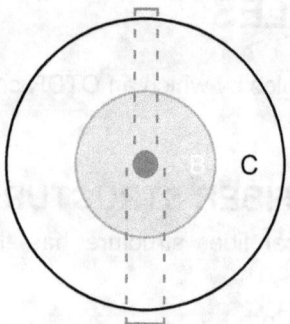

MFD≈ 9.2-11 µm
B=cladding= 125 µm
C= primary coating

Figure 1-3: Singlemode Fiber

First, by surrounding the core, the cladding confines the light to the core. Second, the cladding increases the size of the fiber so that it has increased strength and ease of handling.

The difference in chemical compositions of the core and cladding results in confinement of light in the core. The cladding confines the light by two mechanisms: reflection and wave guiding. In multimode fibers, the light reflects at the core boundary, as long as the angle of incidence is less than the critical angle. The numerical aperture (NA) of the fiber defines this angle. In singlemode fibers, wave guiding confines the light to the core.

The primary coating protects the fiber from mechanical and chemical attack. This protection enables the fiber to retain its intrinsic high strength during cable manufacturing and for a cable lifetime of over 20 years.

1.1.3 WAVELENGTH

Optical transmission systems operate with light. Light has the characteristic of wavelength. Wavelength is a measure of the 'color' of light in nm. The OTDR operator tests at the wavelength at which the network will operate. Table 1-1 presents wavelengths common to fiber optic systems.

	Multimode	Singlemode
System	nm	nm
Data	850, 780 1300	1310
CATV		1310, 1550
Telephone		1310, 1550
DWDM CWDM		1310-1550 1625
WDM	850, 1300	1310, 1550
FTTH/PON		1310, 1490 1550

Table 1-1: Common Communication Wavelengths

1.1.4 ATTENUATION

As light travels along a fiber, its power level drops. This drop is attenuation. The main cause of attenuation is Rayleigh scattering (PFOI, 3.5.2).

Rayleigh scattering occurs when light is scattered by atoms in the core towards the core boundary at an angle greater than the critical angle. Such light escapes from the core (Figure 1-4).

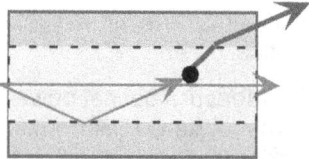

Figure 1-4: Rayleigh Scattering

This scattering is analogous to standing in a dark and dusty room with a high intensity flashlight. As you shine the light, the dust particles are visible from the side, because the dust scatters the light towards the side of the beam.

The OTDR operator is concerned with the attenuation rate, in dB/km. The OTDR provides measurement of this rate. The operator determines whether or not the measured rate is acceptable. To determine acceptability (5.1), the operator needs to know both the maximum and typical rates. Table 1-2 and Table 1-3 present these rates. As these tables demonstrate, the attenuation rate depends on the wavelength of the light and the core diameter of the fiber.

Wavelength, nm	Core Diameter, μ	Attenuation Rate, dB/km
850	62.5	3.5
850	50	3.5
1300	62.5	1.5
1300	50	1.5
1310[1]	8.2	0.5
1310[2]	8.2	1.0
15502	8.2	0.5
1550[2]	8.2	1.0

Table 1-2: Maximum Attenuation Rates

Wavelength, nm	Core Diameter, μ	Attenuation Rate, dB/km
850	62.5	2.8-3.0
850	50	2.5-2.7
1300	62.5	0.7
1300	50	0.7
1310	8.2	0.30-0.35
1550	8.2	0.20

Table 1-3: Typical Attenuation Rates

1.1.5 BACKSCATTER

Some of the light is scattered backwards within the critical angle. Such light travels back to the input end (Figure 1-5). All OTDR measurements are of power that returns to the OTDR.

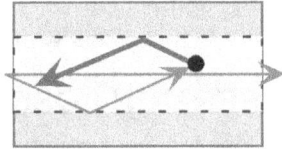

Figure 1-5: Backscattered Power

This backscattering is analogous to being in a dark dusty room with a high intensity flashlight. You can see the suspended dust because it scatters light backwards towards you.

1.1.6 BACKSCATTER COEFFICIENT

The OTDR operator calibrates the OTDR to the fiber under test. The backscatter coefficient is the first parameter that provides this calibration. With an incorrect coefficient, the attenuation rate values will

not be accurate. Table 1-4 and Table 1-5 provide common backscatter coefficients.

Wavelength=	850	1300
Corning Clearcurve	-68.0	-76.0
OFS	-68.4	-75.8

Table 1-4: Multimode Backscatter Coefficients

Wavelength=	1310	1550	1625
Corning SMF28e+	-77.0	-82.0	
LEAF		-81.0	-82.0
Draka/ G.652	-79.4	-81.7	-82.5
Draka Teralight/ G.655	-77.4	-80.4	-81.3

Table 1-5: Singlemode Backscatter Coefficients

1.1.7 INDEX OF REFRACTION

As stated earlier, the OTDR launches a pulse of light into the fiber and measures round trip travel time of the pulse and the power level of the returning pulse. This round trip travel time is of no use to the OTDR operator. He needs to know the location of power loss along the length of the fiber.

To calculate this location, the operator inputs the index of refraction (IR) of the fiber under test. This IR is a measure of the speed of light in the fiber. With this index, the OTDR converts the round trip travel time to a distance. In effect, the index of refraction is the second parameter that calibrates the OTDR to the fiber under test,

The index of refraction is defined as the ratio of the speed of light in a vacuum to that in the fiber. As a ratio, the IR has no units of measure. The IR of optical fibers ranges from 1.46 to 1.52 (Appendix 8.1). Error in the IR can result in distance errors that range from less than a meter to more than 47 m (Appendix 8.3).

1.1.8 FIBER LENGTH VS. CABLE LENGTH

The index of refraction calibrates the OTDR so that it provides accurate fiber length. However, fiber length and cable length are different.

---

[1] Loose tube cable

[2] Tight tube cable

There are two causes of this difference. All cables enclose fibers in buffer tubes. The first cause is a buffer tube that spirals around the cable axis. This spiraling results in buffer tube length that is longer than cable length. In loose tube cables, the fiber length is longer than the buffer tube. The sum of these two factors can result in fiber length that is 3% more than cable length. Appendix 8-2 presents examples of this difference.

1.1.9 REFLECTIONS

The power returning to the OTDR has two sources. The first is backscatter, as presented. The second is reflected power.

Almost all fiber ends create a reflection that sends power backwards to the OTDR. This reflected power results from a change in the index of refraction. Such a change occurs at fiber ends.

This change in the index of refraction (IR) creates a 'Fresnel' reflection. A common example of a Fresnel reflection occurs when looking through a closed window. While you see through the window, you may see a dim reflection of yourself or the room behind you. This dim reflection is a Fresnel reflection.

Reflections occur at three types of fiber ends:

 ➢ Non contact connectors
 ➢ Contact connectors
 ➢ Fiber breaks

Non-contact connectors have an air gap between ends. The difference between the IR of the fiber, 1.46-1.52, and that of air, approximately 1.0, results in a reflection.

Contact connectors make physical contact. However, these connectors have microscopic air gaps between ends. These gaps result from imperfectly smooth fiber ends. The difference between the IR of the fiber and that of air results in a reflection. Since these air gaps represent a smaller fraction of the cross section of the fiber than do non-contact connectors, the reflectance of contact connectors is lower than that of non-contact connectors.

Tight tube cable designs are also known as premises, distribution, and tight pack, designs. Fiber breaks in tight tube designs create fiber ends. The tight buffer tube can keep these ends in alignment and in contact. In this case, the optical signal path is continuous. However, the lack of perfect smoothness of the break creates a location of change in IR. This change creates a reflection.

An APC connector creates a fiber end. As such, it might create a reflection. However, this green, singlemode connector has an 8° angle on the end face. This angle reflects light backwards outside of the critical angle of the fiber. Such light is lost and does not return to the OTDR. Thus, APC connectors do not create a reflection.

1.2 REASON FOR USE

Installers perform power loss tests on installed fiber links. Such tests have two prime functions:

 ➢ Verification that the link loss is low enough for proper transmitter and receiver operation
 ➢ Verification that the loss of each cable segment, splice, and connector pair is acceptable

1.2.1 INSERTION LOSS TEST

Two power loss tests are common:

 ➢ Optical loss, or insertion loss
 ➢ OTDR

The optical loss or insertion loss, test simulates, somewhat imperfectly, the operation of a transmitter receiver pair (Figure 1-1). This test is a measure of the difference between the power level delivered to the input of a link (i.e., at the transmitter) and the level that exits the output end (i.e., at the receiver). This test provides a measurement of the total loss of the link.

This test is simple to make and to interpret. In addition, this test requires relatively low cost equipment.

However, the location and distribution of the loss are unknown. As a result, conditions of reduced reliability can result in an acceptable insertion loss. For example, an acceptable, low insertion loss can result from at least four situations:

> Nominal loss connectors and nominal attenuation rate cable

> Low loss connectors and high attenuation rate cable

> High loss connectors and low attenuation rate cable

> Low loss connectors, low attenuation rate cable, and a bend radius violation, or some other violation of the cable performance parameters

Since the insertion loss test is blind to the location and distribution of power loss, it cannot be used to indicate proper installation of all components of the link. By 'component', we mean a cable segment, splice, or connector pair.

At best, the insertion loss test provides a crude inference of proper installation. In a strictly technical and legal sense, a low insertion loss measurement does not *prove* low loss or proper installation of all components in a link. To provide such an indication, another test is needed. That test is optical time domain reflectometry (OTDR).

1.2.2 OTDR TEST

The OTDR provides loss data on the location and distribution of power loss along a fiber optic link. These data reveal the loss of almost all link components. With such testing, the operator can verify proper installation of each component. If each component in a link is properly installed, the link will have the maximum reliability possible.

> This, maximum reliability, is the key advantage of, and reason for, OTDR testing.

From OTDR test results, an operator can compare measured values for each component to acceptance values. This comparison reveals the reliability with which each component has been installed. No other test can provide this indication of reliability.

1.3 LOSS SIMULATION

For many years, some professionals have used the OTDR to perform a test that attempts to simulate an insertion loss test by including the loss of connectors at both ends. Moreover, the data standards approve such a test. The Building Wiring Standard, TIA/EIA-568-C, references 61280-4-1, which allows use of OTDR testing.

Such testing has two problems. The first problem is that the OTDR can exhibit positive losses, known as 'gainers' (3-1 and Figure 3-2) These positive losses result in an underestimate of both the insertion loss and the loss that a transmitter-receiver pair will experience.

The second problem is that the OTDR signal does not travel in same manner as that of an optical signal. In a fiber, the optical signal travels in one direction; in an OTDR test, the signal travels in both directions.

These two problems do not provide technical support for use of an OTDR to make the equivalent of an insertion loss measurement. This author's experience is that an OTDR test that 'simulates' an insertion loss test produces a value that is lower than that of a true insertion loss test.

1.4 TYPES

There are three types of OTDRs:

> Mainframe OTDRs (Figure 1-6)

> Mini- OTDRs (Figure 1-7 and Figure 1-8)

> OTDR modules (Figure 1-9)

The AC-powered mainframe OTDRs, with high-speed processors, are used in fiber and cable manufacturing facilities. The high speed justifies their high cost.

A portable, battery powered, mini-OTDR meets the needs of most field installation testing. This OTDR has a reduced-speed processor for extended operating time.

OTDR modules are cost effective when the purchaser already has a computer or notebook computer than can be used as an OTDR. Modules are controlled and powered by an external computer, such as a notebook or netbook computer (Figure 1-9). The module has the two additional advantages of lightweight and small size.

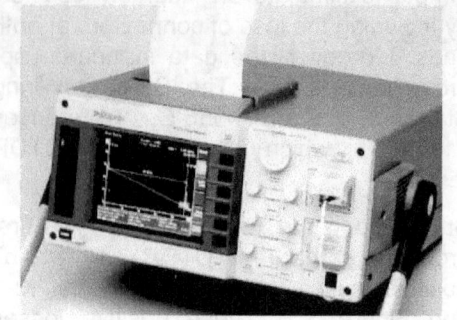

Figure 1-6: Mainframe OTDR (Tektronix)

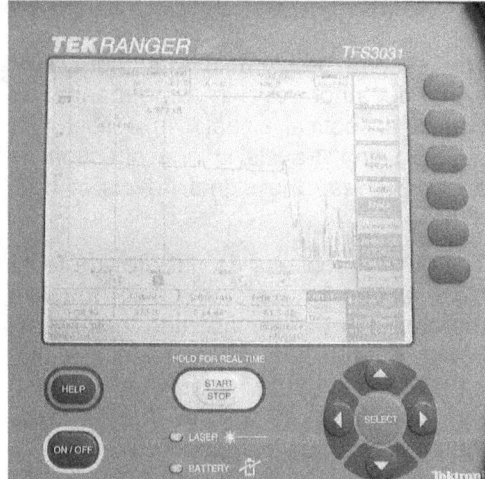

Figure 1-7: Tektronix Mini-OTDR

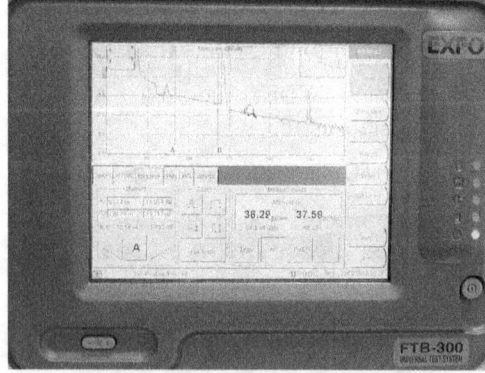

Figure 1-8: EXFO Mini-OTDR

Figure 1-9: USB-Powered OTDR Module

1.5 BLOCK DIAGRAM

In its simplest description, an OTDR consists of four parts (Figure 1-10):

➤ A high-power laser diode

➤ A high sensitivity detector

➤ A bi-directional coupler

➤ A connector on the front panel

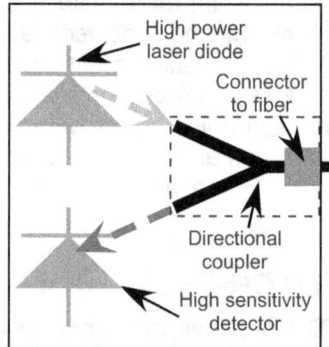

Figure 1-10: Functional Diagram Of OTDR

The laser diode sends a pulse of light into a fiber. At each point along the fiber, some of the light is scattered or reflected backwards to the OTDR. The directional coupler directs this power to the high sensitivity detector. The OTDR measures received power level, in dB, as a function of round trip travel time.

1.6 THEORETICAL TRACE

When the index of refraction (PFOI, 2.2.3) for the fiber is entered into the OTDR, the OTDR displays optical power level as a function of fiber distance. The OTDR displays this information in the form of a 'backscatter' trace (Figure 1-11).

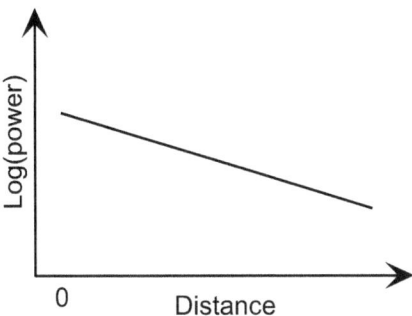

Figure 1-11: Theoretical Backscatter Trace

An important characteristic of this trace is the straight line. The straight line indicates uniform attenuation along the fiber. A properly designed, manufactured, and installed cable segment always has uniform attenuation and a straight-line trace. A deviation from the straight-line trace indicates a problem at the location of the deviation.

> The OTDR operator expects to see a straight line trace for each cable segment.

> Any deviation in a segment indicates a problem.

The backscattered power level is extremely low. For example, the backscatter coefficient for singlemode fibers is approximately -80 dB. With such a value, a 1 mW (10^{-3} W) power level at any point along the fiber will result in a backscattered power level of 10^{-11} W, or one-hundredth of a nanowatt.

Because this power level is very low, it must be amplified. Such amplification creates significant noise. To reduce this noise, the OTDR sends thousands of pulses into the cable. The OTDR averages the returned power measurement from all pulses. With averaging, the noise in the signal is reduced to a level that allows accurate loss measurements.

> The OTDR makes multiple measurements.

> The OTDR operator determines the number of measurements.

1.7 MEASUREMENTS

The OTDR enables the following measurements:

> Attenuation rate

> Connector loss

> Splice loss

> Segment length

> Reflectance (but not always)

By measuring returned power levels at different points along the trace, we can measure the loss of *almost* every component and distance to that component. For example, we can measure the power levels at the beginning and end of a cable segment. The OTDR divides the power loss by the distance between these two locations to determine the attenuation rate, also known as the attenuation coefficient. We can measure the backscattered power levels before and after a connection to determine its power loss.

> Power difference is loss

Finally, we can determine cable segment lengths. With these measurements, we can verify acceptably low power loss for almost all components and conformance of the installed cable distances to the design map. In short, the OTDR gives us the ability to verify proper and reliable installation of the link.

1.8 TRACE FEATURES

In this text, we use the symbols in Figure 1-12.

▭▭	radiussed connector pair
▭	radiussed connector
▭	fusion splice
▭▭	mechanical splice
▭◿	APC connector pair

Figure 1-12: Symbols Used

1.8.1 REFLECTANCE

The trace in Figure 1-11 includes only power back scattered from atoms in the core. However, connectors and splices can reflect power backwards towards the input end (1.1.9). The power from these reflections adds to the backscattered power. When the OTDR adds this reflected power, the straight-line trace has peaks at the locations of reflections. The peaks are also known as spikes, but are most accurately called reflections (Figure 1-13).

> Fiber ends can create reflections

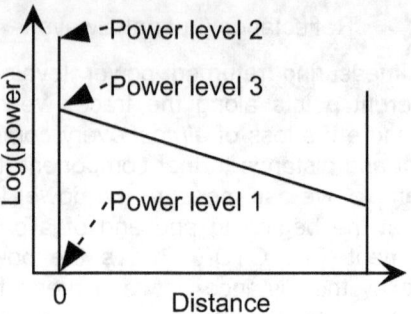

Figure 1-13: Trace With Backscatter And Reflections

The height of a reflectance, or peak, depends on a nature of the connection. The larger the difference in the index of refraction at the connection, the higher will be the reflectance. Reflectances of various connections, from highest to lowest, are:

> Air gap connectors

> Flat physical contact connectors

> Radius physical contact connectors

> Mechanical splice

> Fusion splice and APC connectors

1.8.2 DEAD ZONES

In Figure 1-13, the peaks are vertical lines. These vertical lines imply an OTDR ability to measure three different power levels in zero time. This implication cannot be correct, since all electronics have a response time; that is, a time to respond to changes in power level. The OTDR cannot display a figure like Figure 1-13 because of this response time. Instead of vertical lines, as in Figure 1-13, the trace has peaks with a forward slant and a finite width (Figure 1-14).

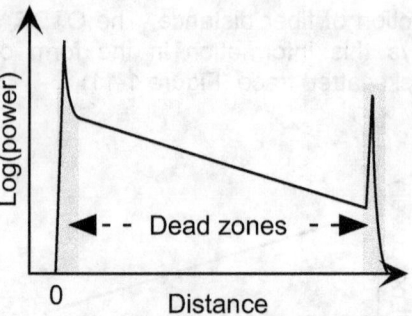

Figure 1-14: Basic Trace With Modified Peaks

The width of these peaks is a characteristic of the OTDR and not of the cable under test. This peak width obscures the characteristics of the fiber in the peak. Because of these facts, we call these widths 'dead zones' or 'blind zones'.

This width of the dead zone can be from a few meters to thousands of meters. This width is determined by two factors:

> Bandwidth limitation

> Overload power level

As the bandwidth of the OTDR detector increases, the width of the dead zone decreases. If the returning power overloads the detector, it the takes time to recover from the overload condition.

> A dead zone, or blind zone, exists at all locations of a power change that occurs at a specific location.

> A power change can occur at reflective and non-reflective events.

> Peaks can conceal features

1.8.3 CONCEALED FEATURES

Concealed features, i.e., reflectances or power drops, occur whenever the features are more closely spaced than the width of the dead zone.

Imagine there are two reflective components, such as contact connectors, that are closely spaced (Figure 1-15). The peak created by the first reflective component can be wider than the distance between the two components. In this case, the peak from the first component can conceal the peak from the second component (Figure 1-15).

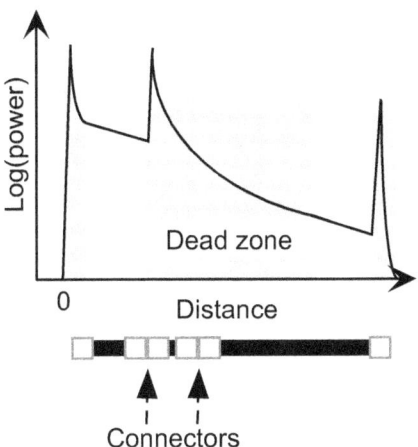

Figure 1-15: Two Closely-Spaced Reflective Events

Because a reflection from one reflective component can obscure the presence of another, closely spaced, component, we may not be able to measure the loss of every component in a link. Instead, we can measure the total loss of multiple components. Thus, we may not be able to test and confirm proper installation of all components. Hence, in 1.1, we used the phrase 'testing of *almost* every component in a fiber optic link'.

1.8.4 MAP TO TRACE

Because we do not know what is in the dead zone, we cannot create a map from a trace. Instead, we must have an accurate link map in order to interpret a trace properly. Figure 1-15 provides an example: two pairs of connectors (Figure 1-15) can create essentially the same trace as would a single pair of connectors (Figure 1-16). In summary:

➢ You cannot create an accurate map from a trace.

➢ You can only interpret the trace from an accurate map.

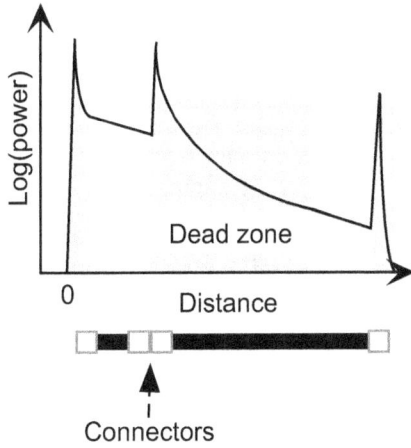

Figure 1-16: Trace Of Connectors, Single Pair

1.9 MEASURING REFLECTANCE

Reflectances have two types of tops: sharp and flat. A sharp top indicates that the OTDR detector has not been overloaded. In this case, the OTDR can be used to measure reflectance of connectors and mechanical splices.

A flat top indicates that the OTDR detector has been overloaded. In this case, the OTDR cannot be used to measure reflectance accurately. When the detector is overloaded, the reflectance value will be an underestimate of the real value. In other words, a reflectance measurement of -35 dB could actually be -25 dB. This relationship is the opposite of what we would prefer.

FOTP-8, also know as EIA/TIA-455-8, is a test standard that describes reflectance testing with an OTDR. This standard limits use of the method to traces that have sharp tops (i.e., not overloaded). This author's experience is that most OTDRs exhibit flat tops. Such OTDRs will provide non-conservative reflectance measurements.

1.10 KEY CONCEPTS

1. For low loss, the core in the launch cable must match that in the cable under test.

2. Singlemode cores are much smaller than multimode cores.

3. OTDR testing is performed at wavelength(s) of the transmitter(s).

4. The OTDR operator measures the attenuation rate, in dB/km, of each cable segment.

5. The IR and backscatter coefficient calibrate the OTDR to the fiber under test.

6. Pulses traveling forward in the fiber create power scattered backwards.

7. Cable length is less than fiber length.

8. The OTDR test provides information on the reliability of the components of the link.

9. Each cable segment exhibits a straight-line trace.

10. Deviations from a straight-line trace for a cable segment indicate a problem at the location of the deviation.

11. The OTDR creates a trace from the average of multiple measurements.

12. A change in IR creates a partial reflection, also known as a Fresnel reflection.

13. Reflections occur at fiber ends.

14. Dead zones conceal features.

15. Dead zone occur at both reflective and non-reflective events.

16. You cannot create an accurate map from a trace.

17. You can only interpret the trace from an accurate map.

2 BASIC TRACES

Chapter Objectives: you learn the three basic traces and their potential origins.

2.1 TRACE TYPES

With a simple understanding of the OTDR and its traces, we can reduce thousands of traces to three basic traces:

➤ A reflective loss

➤ A non-reflective loss

➤ A bad launch

All other traces are combinations of these three basic traces.

2.2 REFLECTIVE LOSS

A reflective loss will have the appearance of Figure 2-1. This trace can result from at least five link configurations:

➤ Two segments connected by radius connectors

➤ Two multimode segments connected by mechanical splice

➤ Two singlemode segments connected by a mechanical splice

➤ A broken fiber in a tight tube cable

➤ A single segment with multiple reflections

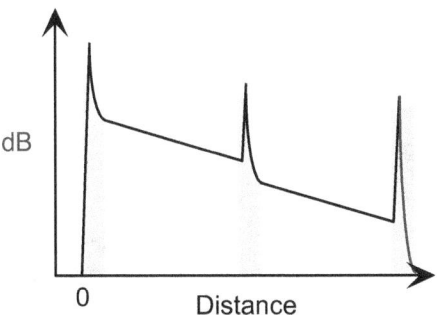

Figure 2-1: Reflective Loss

2.2.1 RADIUS CONNECTORS

At the time of this writing, all connectors used in data, telephone, and CATV networks are either radius or APC connectors. Radius connectors have a radius of curvature on the connector end. These connectors make physical contact. The OTDR trace of two cable segments connected by radius connectors will exhibit reflectance at the location of the connectors (Figure 2-2). This reflectance results from the imperfectly smooth surfaces of the connectors (PFOI, 5.4.5). This lack of perfect smoothness creates microscopic air gaps. At these gaps, there is a change in index of refraction [IR].

➤ Radius connectors always create reflectance.

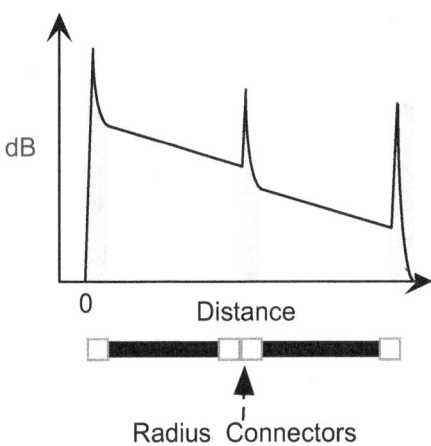

Figure 2-2: Radius Connectors Trace

2.2.2 MULTIMODE MECHANICAL SPLICES

Multimode mechanical splices have an index matching gel between the fiber ends. Reflectance results from the difference between the IR of the gel in the splice and the IRs of the cores of the graded index (GI) fibers. Since the multimode cores have multiple indices of refraction, the single IR of the gel will be mismatched to most of the core.

➤ Multimode mechanical splices always create reflectance (Figure 2-3).

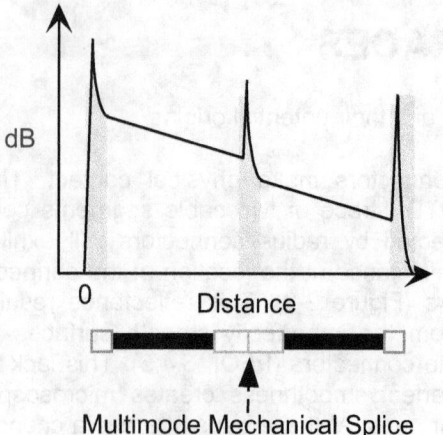

Figure 2-3: Multimode Mechanical Splice Trace

2.2.3 SINGLEMODE MECHANICAL SPLICES

Singlemode mechanical splices contain two fibers and an index matching gel. Reflectance occurs if the IRs of the two fibers and that of the gel are not the same.

➤ Singlemode mechanical splices may create reflectance (Figure 2-4).

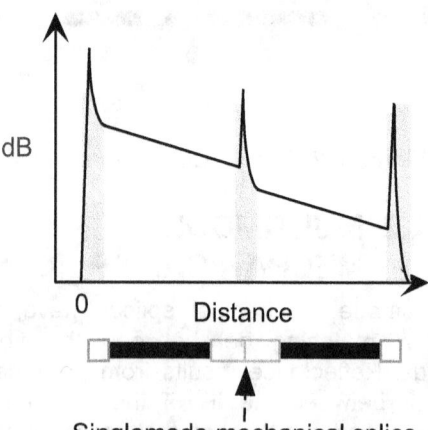

Figure 2-4: A Singlemode Mechanical Splice Trace

2.2.4 BROKEN FIBER

In a loose tube cable, broken fiber ends fall apart. In a tight buffer tube cable, the buffer tube can align the ends of a broken fiber. A broken fiber in a tight tube cable creates reflectance. The OTDR trace of such a break exhibits a reflectance at the

break (Figure 2-5). The cause of this reflectance is air between the fiber ends.

➤ A broken in fiber in a tight tube cable always creates reflectance.

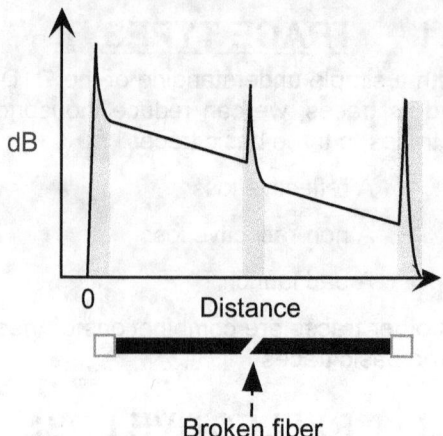

Figure 2-5: Broken Tight Tube Cable Trace

2.2.5 MULTIPLE REFLECTION

A single segment of cable with a high reflectance connector may create multiple reflectances (Figure 2-6). Light scattered backwards from the core and reflected by the connectors can be partially reflected back into the fiber at the OTDR. This reflected power makes a second round trip, before entering the OTDR and registering as a peak (Right peak of Figure 2-6).

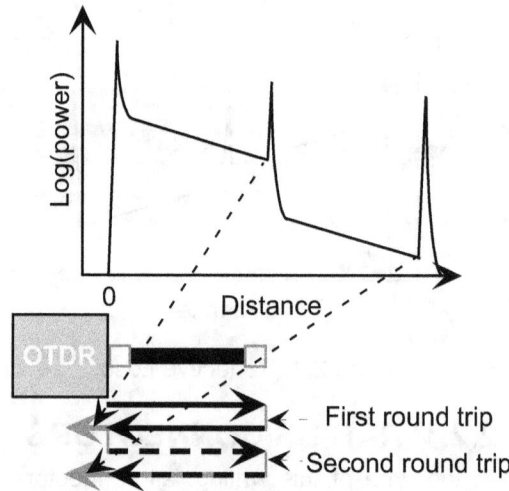

Figure 2-6: Reflective Event From Multiple Reflection

This reflection is a 'multiple reflection', commonly known as a 'ghost reflection'. However, the power from the second round trip arrives after the power from the first round trip.

> 'A single segment may create two apparent segments.

The operator must identify multiple reflections as such in order to justify ignoring them Such identification is necessary, as the multiple reflection may be a broken fiber in tight tube cable.

The operator can identify ghost reflections from three characteristics:

> Length

> Height of second peak

> Trace from opposite end of cable

Imagine a single segment. Without a ghost, we would expect to see a trace like that in Figure 2-7. If we see a second segment (Figure 2-8), we would expect the second segment to have almost exactly the same length as the first.

If the ghost appears at the end of a single segment, as shown, the multiple reflection will appear at exactly twice the distance to the first. However, if the multiple reflection appears in the middle of a subsequent segment, the distance to the second peak is at almost, but not exactly twice the distance to the first. As a result of the forward slant of the peak, the second peak will appear at slightly more than twice the distance to the first.

As the power entering the OTDR after the second round trip is lower than that entering after the first round trip, we would expect the height of the second peak to be lower than that of the first. This is the case.

Ghosts can appear on a trace from one end of the cable, but not from the other. In this case, the reflection is obviously a ghost, as reflections are from fiber ends. A real fiber end in the cable will create a reflection in both directions.

When the ghost does not appear from both ends, the connector on the end from which the ghost appears usually has a loss higher than that of the connector on

the opposite end. The reason for this is simple: as a ghost reflection has a power level lower than that of a real reflection, the closer to the OTDR noise floor the signal level is, the more frequently a ghost reflection appears.

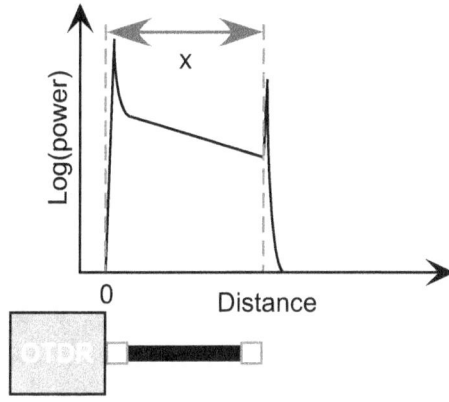

Figure 2-7: Single Segment Trace Expected

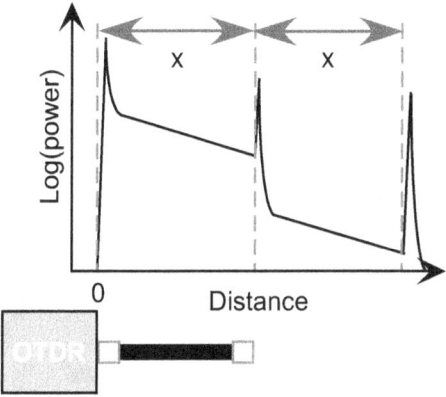

Figure 2-8: Ghost Segment After Segment

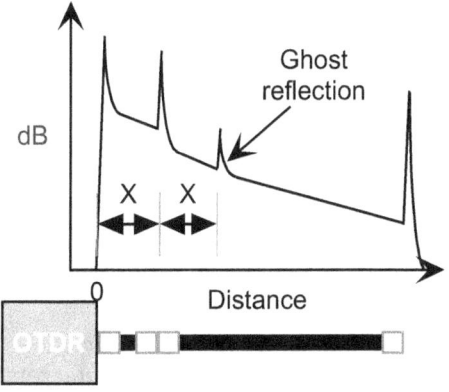

Figure 2-9: Ghost Reflection In Middle Of Segment

One situation in which a ghost reflection is most troubling is a multi-segment cable with a short segment followed by a long segment. In this case, a ghost reflection from the short segment can appear in the middle of the long segment (Figure 2-9). The operator must identify the origin of this reflection, as it could be a broken fiber.

To determine whether the unknown peak is a ghost reflection, the operator tests from both ends. When the cable is tested from the near end, the unknown reflection occurs at about 2X from the near end. If the reflection is a real fiber end and not a ghost, the reflection will occur at a distance of 1x from the far end when the cable is tested from the second end (Figure 2-10). If the unknown reflection occurs at 2X from the OTDR at both ends, the reflection is a ghost. If the reflection does not appear in both directions, the reflection is a ghost.

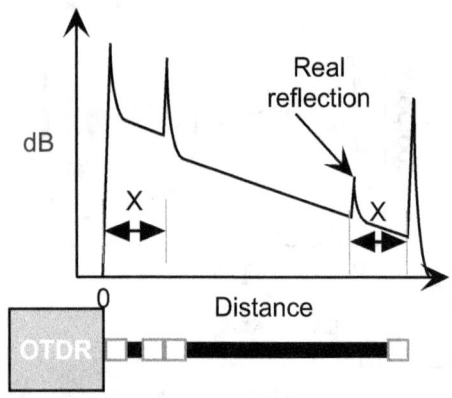

Figure 2-10: Real Reflection

2.2.6 SUMMARY

A comparison of Figure 2-1 through Figure 2-6 reveals that different link components can create nearly identical traces. This comparison reinforces two important principles:

> ➢ Because we do not know what creates the reflection, we cannot create a map from a trace.

> ➢ Instead, we must have an accurate link map to properly interpret a trace.

2.3 NON-REFLECTIVE LOSS

The second basic trace is a non-reflective loss (Figure 2-11). This trace can result from at least four cable configurations:

> ➢ Two segments connected with a fusion splice

> ➢ Two singlemode segments connected with a mechanical splice

> ➢ Two segments connected by APC connectors

> ➢ A cable with a violation of any of its performance parameters

2.3.1 FUSION SPLICE

A properly made fusion splice of the same fiber type creates no reflectance but may exhibit a loss (Figure 2-11). Because there is little or no change in the IR as light moves from one fiber to another, there is no reflectance. Should the fusion splice exhibit a 0.0 dB power loss, the splice would not appear on the trace (Figure 2-12).

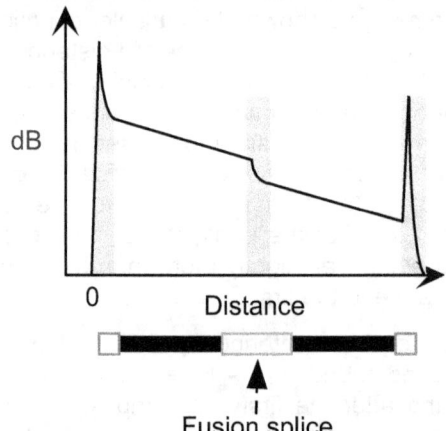

Figure 2-11: Non-Reflective Fusion Splice

To be technically precise, we must state that a properly made fusion splice creates *low* to no reflectance. A small difference in the IRs of the two fibers can create some reflectance. However, in 34 years of viewing traces, this author has never seen a reflection from a properly made fusion splice.

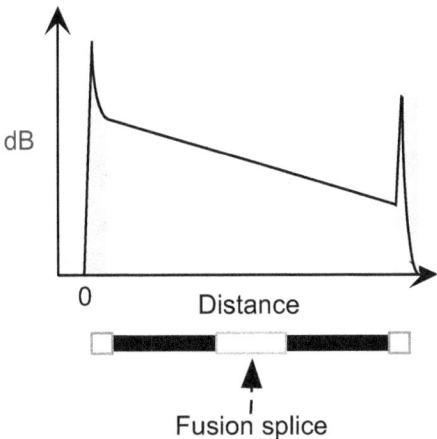

Figure 2-12: Trace For 0 dB Fusion Splice

➢ A properly made fusion splice of either multimode or singlemode fiber never creates reflectance.

Figure 2-12 reinforces the important rule. Because a non-reflective, zero loss splice does not appear on a trace,

➢ We cannot create a map from a trace.

2.3.2 SINGLEMODE MECHANICAL SPLICES

Singlemode mechanical splices may create no reflectance. The core IRs of the singlemode fibers may exactly match the IR of the gel in the splice. Two single-mode segments connected with a mechanical splice may exhibit a loss without a reflectance (Figure 2-13).

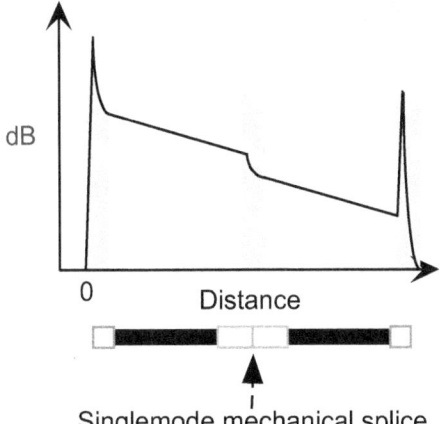

Figure 2-13: Non-Reflective Loss From Singlemode Mechanical Splice

Figure 2-4 and Figure 2-13 may create confusion. Both figures could result from the same cable configuration. If the IR of the gel does not exactly match the IRs of both fiber cores, reflectance will result. As stated earlier, the principle is:

➢ The trace must be interpreted with an accurate map.

2.3.3 APC CONNECTORS

Two segments connected by APC connectors exhibit a loss without reflectance (Figure 2-14). The APC connectors reflect light backwards at an angle greater than the critical angle of the singlemode fiber. When this light crosses the core boundary, it escapes into the cladding. In this case, there will be no reflectance.

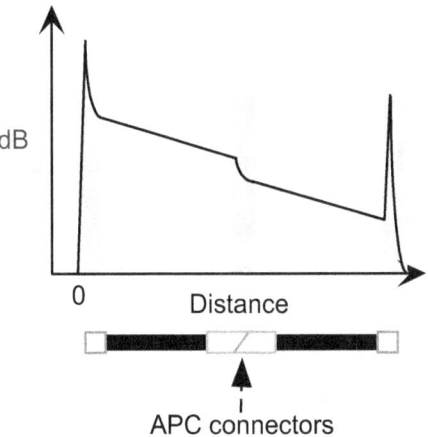

Figure 2-14: Non-Reflective Loss From APC Connectors

➢ Two segments connected by APC connectors never create reflectance.

2.3.4 CABLE PARAMETER VIOLATION

A cable performance parameter violation can be a violation of the bend radius, the crush load, the operating temperature range, the crush load, or any other cable performance parameter. When this violation occurs, power is lost at the location of the violation. We refer to this loss as an 'event' or a 'feature', as it is not a connection. As long as this violation does not cause a broken fiber, it results in an event with a non-reflective loss (Figure 2-15).

➤ Any violation of a cable performance parameter can create loss without reflectance.

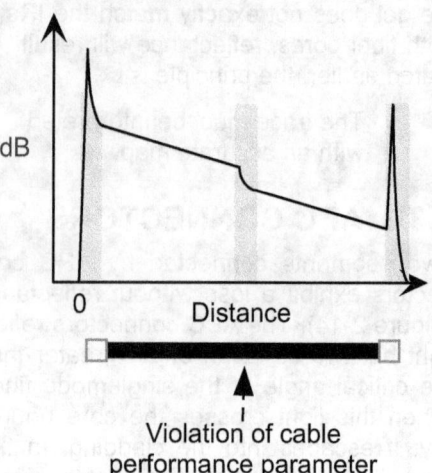

Violation of cable performance parameter

Figure 2-15: Non-Reflective Loss From Cable Parameter Violation

2.4   BAD LAUNCH

The third basic trace is a bad launch (Figure 2-16). A bad launch results from at least three situations:

➤ A broken connector at the OTDR

➤ A break in the fiber within the dead zone

➤ A cable shorter than the width of the dead zone

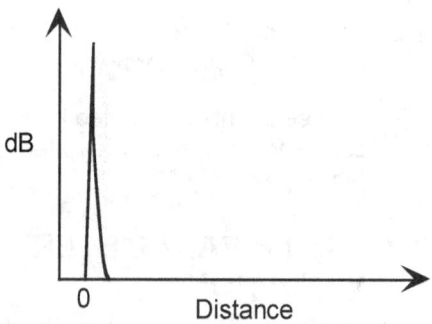

Figure 2-16: Bad Launch

Obviously, if the connector at the OTDR has a broken fiber, no light can enter and return from the fiber. In addition, a broken fiber in the dead zone will be concealed by the reflectance of the connector at the OTDR. Finally, the OTDR cannot display a trace from a cable with a length less than the width of the dead zone.

2.5   KEY CONCEPTS

1. Radius connectors always create reflectance.

2. Multimode mechanical splices always create reflectance.

3. Singlemode mechanical splices may or may not create reflectance.

4. A broken in fiber in a tight tube cable always creates reflectance.

5. Multiple reflections, also known as ghost reflections, can occur.

6. Multiple reflections are not real cable features.

7. When ghost reflections are identified as such, the operator can ignore them.

8. A properly made fusion splice of either multimode or singlemode fiber never creates reflectance.

9. APC connectors never create reflectance.

10. Any violation of a cable performance parameter can create loss without reflectance.

11. The OTDR cannot test a cable shorter than the width of the dead zone.

12. We cannot create a map from a trace.

13. We must have an accurate link map to interpret a trace properly.

3   UNUSUAL TRACES

Chapter Objective: you learn to recognize unusual traces, their causes, and corrective actions.

3.1   INTRODUCTION

In addition to the three basic traces in the previous chapter, there are three unusual traces and an important effect:

> No far end reflection

> Gainers

> High reflectance connectors

> Non linear trace

> Wavelength effects

3.2   NO FAR END REFLECTION

A trace may not have a far end reflection (Figure 3-1). This trace can result from three configurations at the far end:

> A cable with APC connector

> A fiber with a high angle cleave

> A singlemode fiber with a mechanical splice

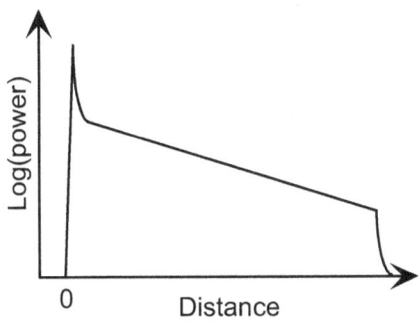

Figure 3-1: Trace Without End Reflection

The APC connector and a high angle cleave reflect light backwards at an angle greater than the critical angle of the fiber. This reflected power escapes from the core and does not return to the OTDR. Thus, no reflection, or peak, results.

A singlemode mechanical splice on the far end may have exact matching of the IRs of the core to that of the gel. With exact matching, no reflection results.

3.3   GAINERS

3.3.1 CAUSE

Previous sections may create the impression that OTDR connection losses are always negative. Such is not the case. It is possible to measure a positive connection loss, also known as a 'gainer' (Figure 3-2). This author has observed gainers on both multimode and singlemode fibers gainers as high as 1 dB. However, most singlemode gainers have been less than 0.5 dB. Since it is not real,

> A gainer must be measured in the opposite direction.

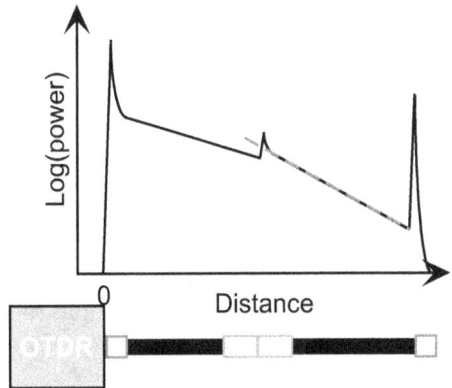

Figure 3-2: Gainer

Gainers are artifacts of OTDR testing. A gainer indicates that more optical power scatters backwards from after the connection than from before. As such, a gainer is not a real gain in power.

A gainer can result from two conditions in the fibers:

> Different attenuation rates

> Different core diameters, or mode field diameters

To understand how different attenuation rates can create a gainer, we return to the cause of attenuation. At each point along its length, a fiber scatters power backwards towards the input end. If we splice two fibers with the same attenuation rate, same core diameter, and same NA with a perfect splice, the splice loss is 0 dB.

However, if the attenuation rate of the fiber before the splice is lower than that after the splice, the power scattered backwards by the fiber before the splice is less than that scattered backwards after the splice. A gain results. However, in the opposite direction, the splice has a loss, because there will be less power scattered backwards from after than before the splice (Figure 3-3).

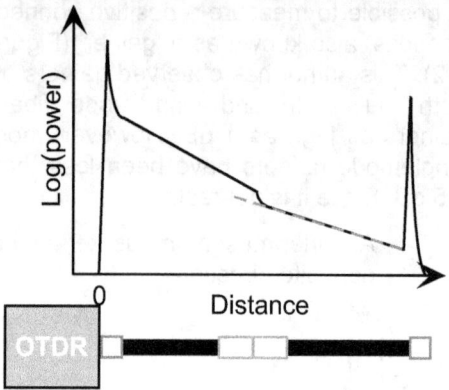

Figure 3-3: Gainer In Reverse Direction

Similarly, differences in multimode core diameters or singlemode mode field diameters can result in gainers. If the multimode core diameter of the fiber before the splice is smaller than that after the splice, there will be more atoms scattering power backwards from after the fiber than before, resulting in a gainer. If measured in the opposite direction, the splice will exhibit a loss.

However, differences in singlemode mode field diameters have the opposite effect: a large MFD before a small MFD produces a gain; in the opposite direction, a loss. Because of these three differences, in attenuation rates, in core diameters, and in mode field diameters, we experience different losses in opposite directions.

3.3.2 TRUE SPLICE LOSS

Since the fibers in a splice can bias the loss high in one direction and low in the opposite direction:

> The true splice loss is the average of the losses in both directions.

Averaging the cancels out the bias.

This principle results in two additional principles:

> A splice with high loss must be tested in the opposite direction.

> A splice with high loss need be remade only if the average loss is unacceptable.

3.3.3 EXAMPLE

Appendix 8.8 presents results of bi-directional OTDR testing of one telephone link. Fifteen splices appear in the first direction. Of these, five splices exceeded the client's acceptance value of 0.25 dB. Eighteen splices appear in the second direction. Of these, five exceed the acceptance value. Although ten splices appear to be unacceptable in one direction, the average loss of all splices was acceptable. This company needed no rework!

In addition, 16 of the 26 splices appeared in one direction only. Since testing in one direction does not reveal all splices,

> Testing in both directions makes sense.

3.3.4 CONNECTOR GAINERS

While this section is on splice gainers, connectors can exhibit gainers also. Since connector loss is higher than splice loss and gainers tend to be small, connectors exhibit gainers less frequently than do splices. The causes of connector gainers are the same as those for splice gainers.

3.4   HIGH REFLECTANCE CONNECTORS

3.4.1 APPEARANCE

High reflectance connectors can cause inaccurate measurements of connector loss and attenuation rate. High reflectance has two appearances:

> Wide dead zone (Figure 3-4)

➢ Overshooting, aka ringing, of the trace (Figure 3-9)

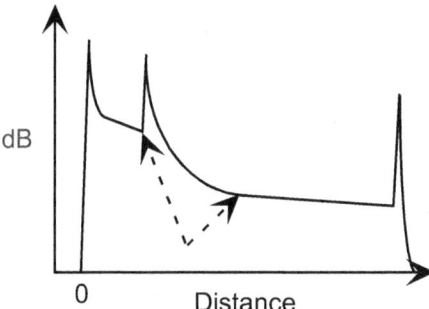

Figure 3-4: High Reflectance Connector

3.4.2 PROBLEMS

High reflectance can overload the OTDR detector, resulting in long recovery time, which is the horizontal axis (Figure 3-4). A wide dead zone can create two problems:

➢ Increased attenuation rate

➢ Increased connector loss

The first problem is increased attenuation rate. With additional power from high reflectance added to the backscatter, the attenuation rate becomes higher than reality. This increase occurs because the automatic trace analysis software performs a LSA on the trace after the reflectance. Some of the reflected power is included in the LSA analysis (Figure 3-5).

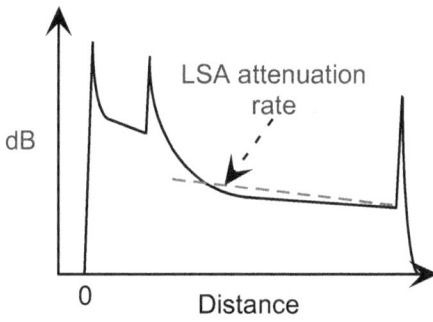

Figure 3-5: Increased Attenuation Rate From High Reflectance

However, when the link is analyzed from the opposite end with a low reflectance connector, the attenuation rate is acceptable. Another method of obtaining accurate rate measurement is cleaning connectors. Cleaning can result in a reduced rate (Figure 3-6). This author has observed the increase to result in incor-

rect rejection of link segments because of increased attenuation rate.

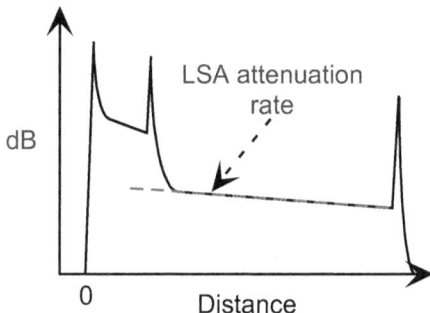

Figure 3-6: Reduced Attenuation Rate With Low Reflectance

The second problem is increased connector loss. Placement of a cursor at the end of the wide reflectance results in increased fiber attenuation included in the connector loss measurement (Figure 3-7). The risk is rejection of low loss connectors (Figure 3-8). This author has observed the opposite: high reflectance connectors that increase the real loss.

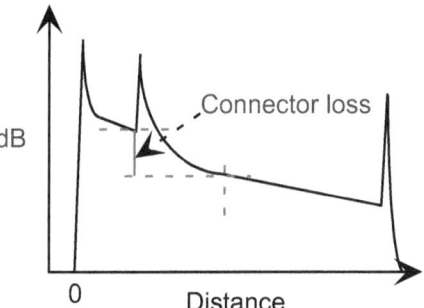

Figure 3-7: Increased Connector Loss From High Reflectance

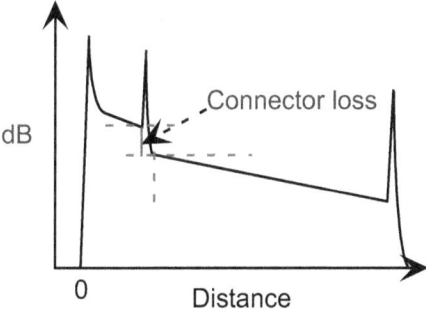

Figure 3-8: Reduced Connector Loss With Cleaned Connectors

High reflectance can result in the trace overshooting, i.e., exhibiting power less than that of the backscatter (Figure 3-9).

As shown in Figure 3-9, overshooting is theoretically impossible. It is impossible for two reasons. The first reason is that the backscattered power cannot exhibit an increase with increasing distance. Because of attenuation, the power must exhibit a decrease in level with increasing distance. The second reason is that it is impossible to have a power level lower than the backscatter level. Such a situation would require cancellation of power. Then, the obvious question is: where did the power go? Answer: overshooting is not real.

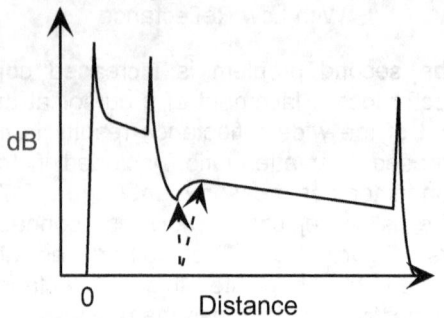

Figure 3-9: Overshooting After High Reflectance Connector

Overshooting can result in three problems:

➢ Reduced attenuation rate

➢ Increased connector loss

➢ Reduced connector loss

If the automatic trace analysis software does not recognize overshooting, the LSA analysis will include the reduced power level in the region of overshooting. This inclusion will result in a reduction of the attenuation slope (Figure 3-10).

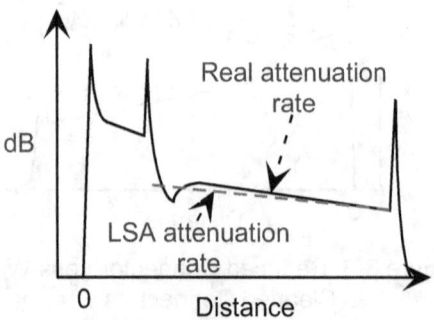

Figure 3-10: Reduced Attenuation Rate With Overshooting

If the OTDR operator chooses the lowest point of the overshooting area, the connector loss will be increased. If the trace analysis software uses the LSA shown in Figure 3-10, the connector loss will be increased. The risk is rejection of a properly installed connector.

Alternatively, the OTDR operator might choose the highest location of the backscatter trace after the connector as a location for connector loss measurement. In this case, the fiber attenuation added to the real connector loss increases. The risk is rejection of a properly installed connector.

This author has two observations on overshooting: it occurs more frequently with short pulse width than with long width; and it occurs on certain OTDRs more frequently than on others.

Consistent overshooting suggests at least three problems:

➢ The OTDR electronics cannot display the high reflectance connectors without overshooting

➢ The launch cable has a high reflectance connector

➢ All the connectors under test have high reflectance

3.4.3 SOLUTIONS

There are four solutions for high reflectance and overshooting.

➢ Connector cleaning

➢ Connector replacement

➢ IR gel application

➢ Pulse width increase

In order to obtain accurate OTDR measurements, the operator cleans or replaces high reflectance connectors. If cleaning or replacement is not practical, application of index matching gel or oil to the connector ferrules can reduce the reflectance and dead zone width. With such reduction, measurements will be accurate. Of course, the gel or oil must be removed from the connectors.

The fourth solution is an increase in the pulse width. This increase results in an increase in the dead zone. If the increase is sufficient, overshooting will not appear in the trace. However, estimated connection loss measurements will increase.

3.5 NON-LINEAR TRACE

In the real world, traces do not always appear as they would in a text. Figure 3-11 and

Figure 3-12 provide an excellent example of this fact and the importance of testing at both wavelengths.

Figure 3-11 shows a fiber that was carrying SONET OC-48 traffic at 1310 nm. The link length was 12,893 m. The link was operating properly. The attenuation rate in each section was uniform, as indicated by the straight-line sections of trace between the splices. The client planned to upgrade the link to SONET OC-192 at 1550 nm. To do so, the client requested 1550 nm qualification testing.

Figure 3-12 shows the 1550 nm trace. Note that the scales of the two traces are different. The reflectance at 0 m is at the OTDR. The second reflectance to the right of the first is the 101 m launch cable. The next drop is at 499 m.

Note the OTDR did not receive a reflectance from the far end at 1550 nm. Note also that there is a drop at about 200 m. This drop does not appear in the 1310 nm trace (Figure 3-11). Finally, note that the trace is not straight, as it should be, were it properly manufactured and installed. This non-linear trace indicates a problem, either with product or installation.

Stress on the fiber was low enough that it did not appear on the 1310 nm trace. However, it was high enough to appear and prevent transmission at 1550 nm.

3.6 WAVELENGTH EFFECT

It is well known (PFOI, 14.3.1.3) that fiber becomes more sensitive to stress as the wavelength increases. As a result of this relationship, comparison of attenuation rates and connector losses at both wavelengths provide an indication of presence or absence of such stress.

Attenuation rates at long wavelength are lower than those at short wavelength [Table 1-2 and Table 1-3].

> If the opposite relationship exists, the fiber is under stress.

Connection losses have two potential relationships at long and short wavelengths. For multimode fibers, connection losses are the same at long and short wavelengths.

> If the loss is higher at the long wavelength than at the short wavelength, the multimode fiber at the connection is under stress.

For singlemode fibers, connection loss can be lower at long wavelength than at short wavelength. This relationship exists because a standard singlemode fiber [G.652, non-dispersion shifted, or 1310 nm fiber] has a larger MFD at the long wavelength than at the short wavelength. As a result, it is possible that the loss will be lower at the longer wavelength.

> If the opposite relationship exists, the fiber at the connection is under stress.

Bend radius violations, either of a cable or a fiber in a splice enclosure, may be evident at a long wavelength but not at a short wavelength. Because of this relationship, comparison of OTDR traces at two wavelengths can differentiate between two conditions:

> A high loss splice that needs to be remade

> A fiber routing problem that can be corrected without splicing

An increased loss at a long wavelength indicates that re-splicing may not be needed. Rerouting fiber and buffer tubes in the enclosure may be sufficient.

In response to this relationship, OTDR manufacturers developed an 'out of band' OTDR, which operates at a wavelength 1625 nm. This wavelength allows full use of the common transmission wavelengths in DWDM networks and the FTTH/PON CWDM networks. An OTDR operating at this wavelength enables simultaneous transmission and monitoring of cable integrity.

3.7 KEY CONCEPTS

1. The far end of a cable may not have a reflection.

2. Gainers exist and are not real.

3. Gainers result from differences between the two fibers at a connection.

4. These differences bias the loss high in one direction and low in the opposite direction.

5. Gainers must be measured in the opposite direction.

6. The true splice loss is the average of the losses in both directions.

7. A splice with high loss must be tested in the opposite direction.

8. A splice with high loss needs to be remade only if the average loss is unacceptable.

9. High reflectance connectors can create errors in attenuation rate and connector loss measurements.

10. High reflectance connectors can create overshooting, which is not real.

11. There are four methods for compensating or correcting high reflectance connectors.

12. Non-linear traces indicate stress on the fiber.

13. Testing at two wavelengths can indicate the type of correction needed.

14. A non-linear trace indicates stress on the fiber.

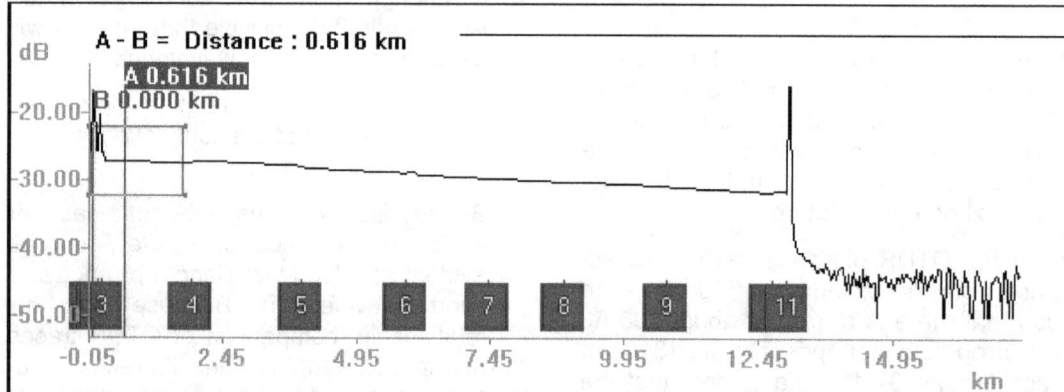

Figure 3-11: Singlemode Link At 1310 nm

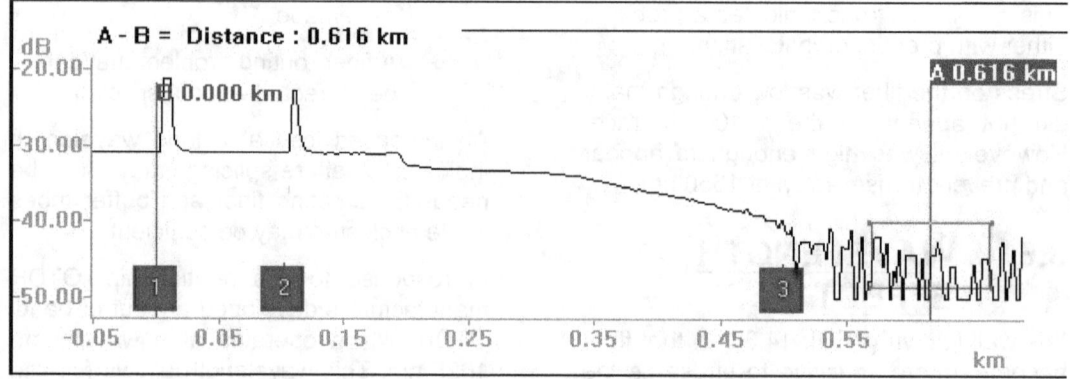

Figure 3-12: The Same Singlemode Link At 1550 nm

4   TESTING AND MEASUREMENTS

Chapter Objective: you learn to set up and make measurements properly.

4.1   SET UP

4.1.1 INFORMATION REQUIRED

The purpose of OTDR testing is measurement of power loss and location of components in a link. In order to create valid measurements, the operator must provide the OTDR with seven types of information:

➢ Wavelength

➢ Pulse width

➢ Maximum cable length to be tested

➢ Either the maximum time allowed for the test or the number of pulses to be analyzed

➢ Index of refraction

➢ Backscatter coefficient

➢ Thresholds

4.1.1.1   WAVELENGTH

The wavelength is the same as that at which the transmitter will operate. If the link will operate with multiple wavelengths, the operator will test at two or more wavelengths. FTTH and DWDM links operate at three or more wavelengths (1310, 1490, 1550 nm). However, testing is done at two (1310, 1550 nm).

Testing at both wavelengths is common and recommended. Unless the operator tests an event at both wavelengths, the true cause of high loss may not be evident (0), Testing at both wavelengths can reduce the cost of correction.

4.1.1.2   PULSE WIDTH

The pulse width determines the power launched into the fiber. As the cable length increases, the pulse width increases to ensure adequate power from the far end of the cable. As the pulse width increases, the dead zone becomes wider. As the dead zone becomes wider, closely spaced events may become concealed.

Testing at two pulse widths may be necessary. Use of a short pulse width reveals events that are close to the OTDR and closely spaced, but may not provide enough power to receive a signal from the far end. Use of a long pulse width provides an adequate signal from the far end.

4.1.1.3   MAXIMUM CABLE LENGTH

The maximum length of cable to be tested controls the time the OTDR takes to complete the test. The operator sets this value above the length of the cable, but as close to that length as possible.

The reason for this setting is simple: the OTDR launches a pulse into the cable and waits until the pulse has had enough time for a round trip. While it waits, it measures the power received. The OTDR repeats this process by the number of pulses or time allowed for the test (1.6).

4.1.1.4   MAXIMUM TIME

The operator uses the OTDR two ways: manual and automatic. In manual mode, the operator specifies the time for the test or the number of pulses to be analyzed. Either determines the noise level in the final trace. The operator learns the appropriate values from experience.

For example, short multimode links (100-1000 m) exhibit relatively noise-free traces after a time of 30-60 seconds or 1024 to 2048 pulses. For singlemode links, a time of at least 60 seconds or 2048 pulses provides relatively noise free traces.

The time allowed or the number of pulses interacts with the pulse width to produce an acceptable trace. For example, an increase in the pulse width enables a reduction in the time or in the number of pulses.

4.1.1.5   INDEX OF REFRACTION

The index of refraction calibrates the OTDR to the fiber being tested. With calibration, the OTDR provides accurate fiber length measurements. However, the fiber length is greater than the cable length. Common values of the difference are in Table 8.1.

Some OTDR software allows entry of a 'helix factor'. This factor is the difference between cable length and fiber length. With a helix factor entered, the distance is cable length. Otherwise, the operator must calculate cable length from the fiber length.

4.1.1.6   BACKSCATTER COEFFICIENT

The backscatter coefficient calibrates the OTDR to the fiber under test. With this calibration, the attenuation rates are accurate. Common values are in Table 8.2.

4.1.1.7   THRESHOLDS

OTDRs allow the setting of thresholds for event loss and reflectance. Setting this threshold removes events that may be caused by noise remaining after signal averaging. Setting thresholds enables reduced time spent on reviewing data that are irrelevant. For example, the operator can set a loss threshold for splice loss of 0.05 dB. With this value, the OTDR will not include events with loss less than this value, simplifying review of the data. Finally, the operator can set a reflectance threshold of -35 or -45 dB, removing reflections less than this value from the event table.

4.1.2 LAUNCH CABLE

To make an OTDR trace, the operator connects a launch cable between the OTDR and the cable under test. The fiber in the launch cable matches the fiber in the cable under test (1.1.2). With a launch cable, the trace will exhibit at least two segments (Figure 4-1).

The launch cable is also known as a 'pulse suppressor'. This label is misleading, in that use of a launch cable was thought to result in a reduction of the dead zone on OTDRs manufactured in the 1980s. However, this author has not

observed such a reduction with OTDRs manufactured after 1992.

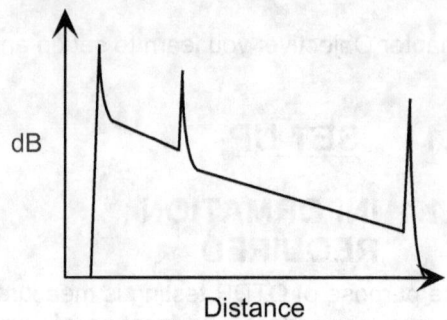

Figure 4-1: Trace With Launch Cable

➤ Use launch cable

4.1.2.1   FUNCTION

The launch cable has two functions:

➤ Protection of the OTDR port

➤ Near end connector loss measurement

When testing a large number of fibers, the operator connects the launch cable to the cables under test. Occasionally, and eventually, the connector on the far end of the launch cable becomes damaged. The repair of the launch cable connector is much less expensive than repair of an OTDR port.

The launch cable enables near end connector loss measurement. Without this launch cable, the loss of the near end connector may not be measurable. To allow this measurement, the launch cable length must be longer than the dead zone.

4.1.2.2   LENGTH

There are two strategies for determining length of the launch cable. The first strategy is a launch cable length longer than the dead zone width. Obviously, the OTDR operator needs to know two values:

➤ Pulse width

➤ The dead zone width of the OTDR at the cable lengths and pulse widths to be tested

With these values, a reasonable minimum launch cable length is three times the length of the dead zone. This length will enable near end connector loss measurement.

The second strategy is to use a launch cable longer than any cable the operator is likely to test. If the launch cable is longer than any cable the operator may test, multiple reflections will appear beyond the end of the cables under test. Thus, multiple reflections will be avoided. Of course, this method of avoiding ghosts is not practical if the cables under test are many miles long.

Multiple reflections may be of no concern if the launch cable connectors and all connectors on the cables under test have low reflectance. This author has performed hundreds of OTDR tests on singlemode telephone cables without any significant number of ghost reflections.

If all connectors do not have low reflectance, multiple reflections will occur. Such reflections need be verified as such so they can be ignored. Of course, bi-directional testing may indicate the truth about multiple reflections (2.2.5). Finally, multiple reflections in loose tube cables are almost always ghosts, as only tight tube cables can have a continuous trace through a broken fiber.

4.1.2.3 REQUIREMENTS

The launch cable should meet six requirements:

> ➢ Matched multimode core diameter, or singlemode mode field diameter,

> ➢ Matched NA

> ➢ Matched connector type

> ➢ Low loss connectors on both ends

> ➢ Low reflectance connectors on both ends

> ➢ Length longer than the dead zone

4.2   MEASUREMENTS

4.2.1 INTRODUCTION

After the operator sets up the OTDR, he obtains a trace of the cable. From this trace, the operator makes loss and length measurements. To make these measurements, the operator can allow the OTDR software to determine all measurements. As we all well know, software is not perfect! Because of this lack of perfection, the operator needs to know how to place cursors on the trace in order to make accurate measurements.

He will do so in two situations. First, he places cursors to verify the accuracy of the software. Second, he places cursors when he prefers manual measurements. In this section, we present the rules for proper cursor placement...so that you need not curse at cursors!

One final note that applies to all manual measurements: the operator makes all measurements with both axes expanded enough to see fine details. If either axis is not sufficiently expanded, the operator may place the cursors incorrectly.

> ➢ Make all manual measurements with scales expanded

The operator makes three measurements:

> ➢ Segment length

> ➢ Splice and connector loss

> ➢ Attenuation rate

4.2.2 LENGTH

Segment length measurements require placement of either one or two cursors. Cursors are placed at the beginning and at the end of the segment.

The beginning of the first segment is at the OTDR, which is at 0 m. For this measurement, the operator need position only one cursor.

The end of any segment is defined by either:

> ➢ A peak for a reflective connection

> ➢ A change in slope or drop off, for a non-reflective connection

For length measurement of the first segment with a reflective connection:

> ➢ The cursor is at the lowest point of the backscatter trace before the peak (Figure 4-2).

For length measurement of the first segment with a non-reflective connection:

> The cursor is at the lowest point of the backscatter trace before the change in slope, or drop off, that marks the end of the first segment (Figure 4-3).

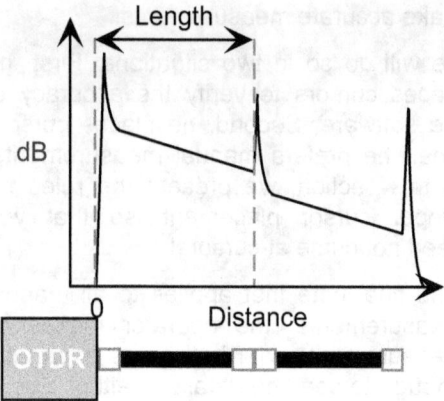

Figure 4-2: First Segment Length Measurement, Reflective Connection

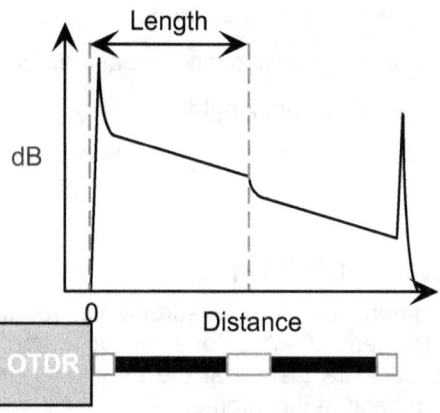

Figure 4-3: First Segment Length Measurement, Non-Reflective Connection

For length measurements of other than the first segment, the operator places two cursors on the trace. The operator places the first cursor at the end of the previous segment and the second cursor at the end of the segment of interest.

For length measurement of a subsequent segment with a reflective connection:

> The first cursor is at the lowest point on the trace before the peak that marks the end of the previous segment (Figure 4-4).

> The second cursor is at the lowest point of a trace of the segment before the peak that marks the end of that segment (Figure 4-4)

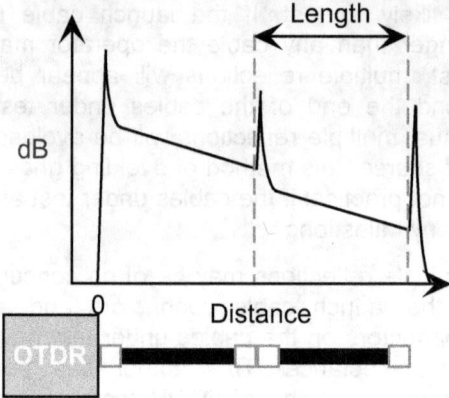

Figure 4-4: Segment Length With Reflective Ends

For length measurement of a subsequent segment with a non-reflective end:

> The first cursor is at the lowest point of the straight-line trace before the drop that marks the end of the previous segment (Figure 4-5).

> The second cursor at the lowest point of a straight-line trace of the segment being measured before the drop that marks the end of that segment (Figure 4-5).

If both ends of the segment are non reflective, the cursors are placed as in Figure 4-6.

In summary:

> A length measurement includes a peak or a drop off at the beginning and excludes a peak or drop off at the end.

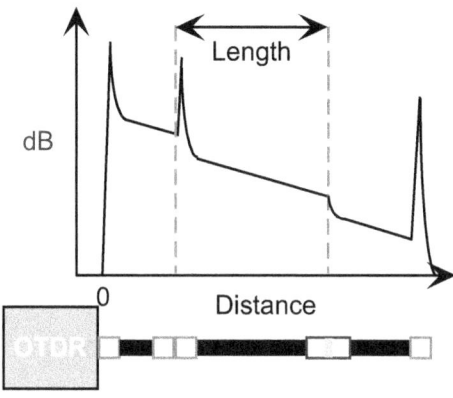

Figure 4-5: Segment With Reflective and Non-Reflective Ends

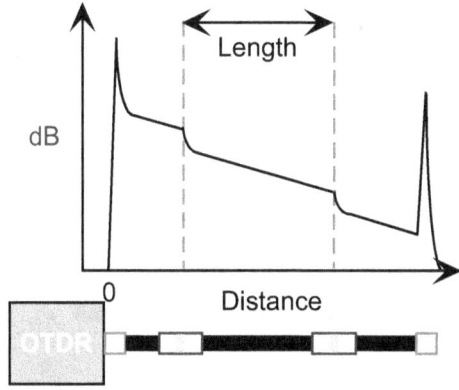

Figure 4-6: Segment Length With Non-Reflective Ends

4.2.3 CONNECTION LOSS

The two types of connection loss measurements are:

> Estimated

> Accurate

The operator makes estimated measurements; the OTDR computer makes accurate measurements.

4.2.3.1 ESTIMATED LOSS

The estimated loss method is also known as the two-point method. The principle for this measurement is:

> Placement of two cursors that straddle the connection, but are in the backscatter, or straight lines, on both sides of the peak or drop.

These measurements overestimate the actual loss. This overestimate results from the fiber attenuation that occurs between the two cursors. As an overestimate, this method is conservative.

For estimated loss measurement of reflective connection:

> Two cursors are placed in the backscatter, or straight-line, trace on both sides of the peak, as close as possible without being in the peak (Figure 4-7).

If the operator places cursors away from the peak (Figure 4-8) the measured loss increases. If the operator places cursors in the peak, the measured loss increases (Figure 4-9) or decreases (Figure 4-10).

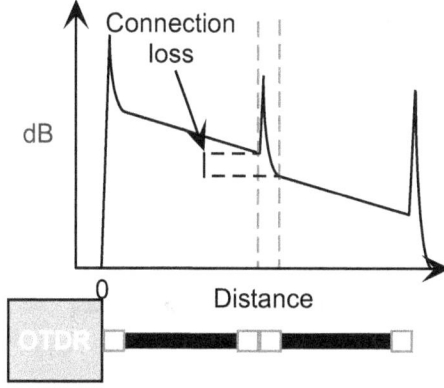

Figure 4-7: Cursor Placement For Estimated Reflective Connection Loss

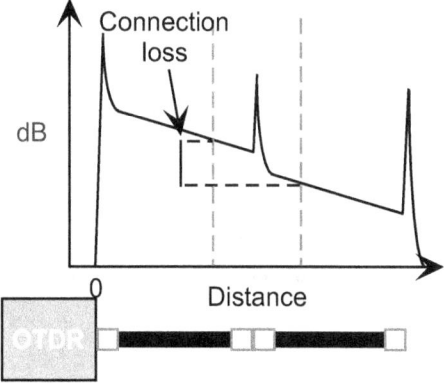

Figure 4-8: Incorrect Cursor Placement 1

For estimated loss measurement of non-reflective connections:

> Two cursors are placed in the backscatter on both sides of the drop off, but as close as possible to the drop off without being in the drop off (Figure 4-11).

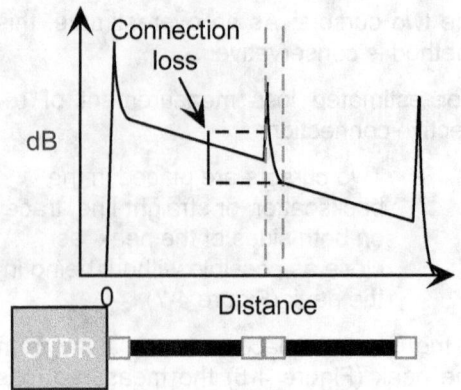

Figure 4-9: Incorrect Cursor Placement 2

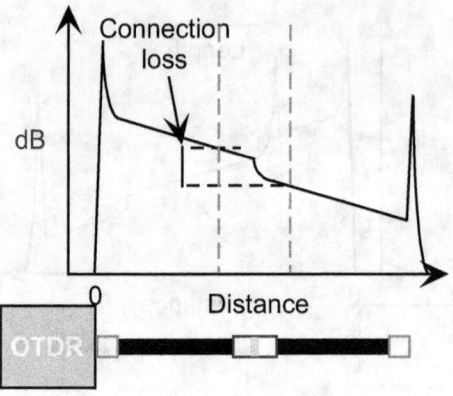

Figure 4-12: Wide Spacing Of Cursors

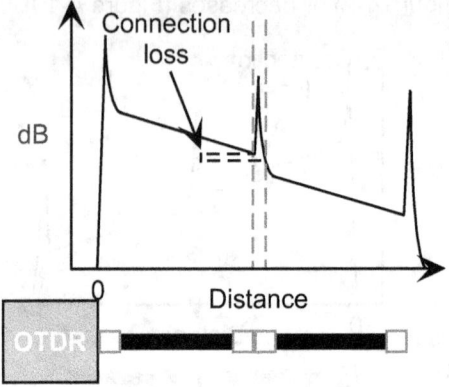

Figure 4-10: Incorrect Cursor Placement 3

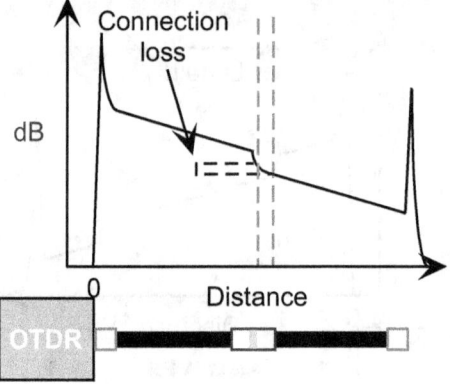

Figure 4-13: Cursors In Drop Off

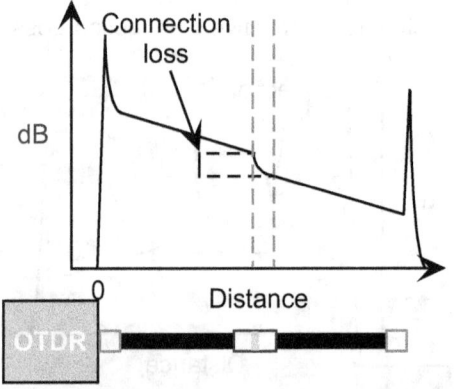

Figure 4-11: Cursor Placement For Estimated Non-Reflective Connection Loss

If the operator places the cursors away from the drop off (Figure 4-12), the measured loss increases. If the operator places the cursors in the drop-off, the measured loss decreases (Figure 4-13).

4.2.3.2 FAR END CONNECTOR LOSS

As indicated, connector loss measurement requires placement of two cursors in the backscatter on both sides of the connection. This requirement cannot be met for the connector on the far end of the link. Measurement of a far end connector loss requires one of two approaches:

➢ The operator moves the OTDR to the far end. With this movement, the far end becomes the near end on the trace.

➢ The operator installs a second 'launch' cable on the far end.

This second cable is a 'receive' cable, similar to the 'receive' test cable in insertion loss testing. With this additional cable (Figure 4-14), the trace meets the placement requirement. Thus, it is possible, though not convenient, to measure the loss of the far end connector from the near end.

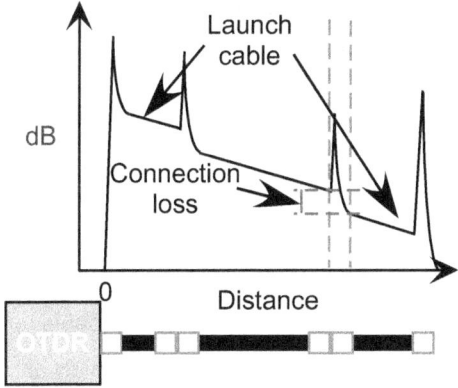

Figure 4-14: Loss Measurement Of Far End Connector

4.2.3.3   ACCURATE LOSS

The OTDR makes automatic accurate loss measurements, also called 'splice loss' measurements,. When this measurement method is active, the computer in the OTDR performs a 'least squares analysis' (LSA) of the straight-line trace that follows the connection. The LSA is also known as a 'least squares fit', The LSA analysis results in a linear equation for the backscatter line after the connection.

The computer extrapolates the line defined by the equation back to the end of the previous segment. The computer calculates the power drop from the end of the previous segment to the extrapolation of the subsequent segment (Figure 4-15 and Figure 4-16). As this method includes no fiber attenuation in the connector loss value, it provides accurate loss values.

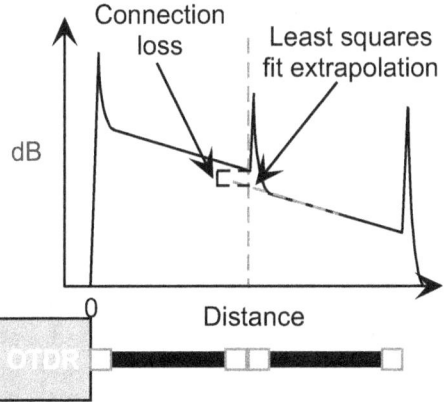

Figure 4-15: Accurate Reflective Loss Measurement

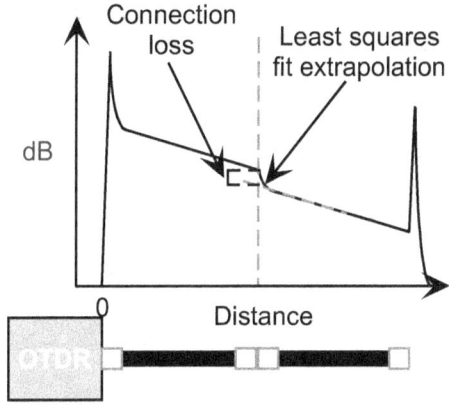

Figure 4-16: Accurate Non-Reflective Loss Measurement

This method requires that the operator place the cursor near the connection. The computer in the OTDR performs the calculation without precise cursor placement.

4.2.4 ATTENUATION RATE

Attenuation rate measurements require two cursors: one each at the beginning and end of the segment. The two requirements for attenuation rate measurements are:

> Placement of two cursors as far apart as possible in the *same* straight-line segment (Figure 4-17)

> Cursors must not enclose any features, such as a drop (Figure 4-18), or a peak (Figure 4-19).

Wide placement creates a value that is most representative of the fiber. Exclusion of drops and peaks results in accurate values.

If the backscatter trace is not straight, the attenuation rate measurement is biased higher than the actual value. In this case, the operator can make attenuation rate measurements on each side of the non-uniform loss (Figure 4-20). However, this is rarely necessary.

A common error is placement of cursors in the peaks or drops that define the ends of the segment. To avoid making this error, the operator moves the two cursors slightly away from peaks or drop offs to ensure that the cursors are properly placed in the trace (Figure 4-21).

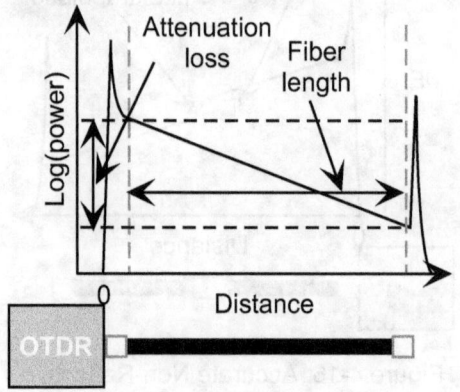

Figure 4-17: Attenuation Rate Measurement

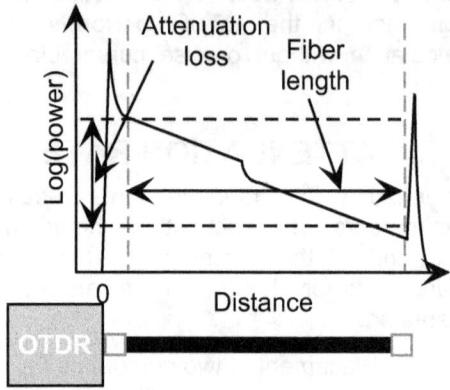

Figure 4-18: Improper Cursor Placement

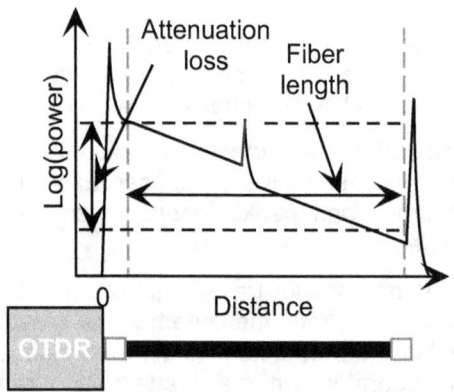

Figure 4-19: Improper Cursor Placement For Attenuation Rate Measurement

Moving the cursors slightly away from peaks or drop offs changes the attenuation rate value slightly. However, the interpretation of the attenuation rate value will not change. That is, measurements made with both cursor placements will indicate

the same condition, i.e., an acceptable rate or an unacceptable rate.

With the information provided thus far, you can determine whether a high attenuation rate is due to defective pro-duct or defective installation. The principles for such determination are:

➤ If the attenuation rate is high and the trace is straight, i.e., uniform, the problem is defective cable.

➤ If the attenuation rate is high and the trace is not straight (Figure 4-18), the problem is defective installation.

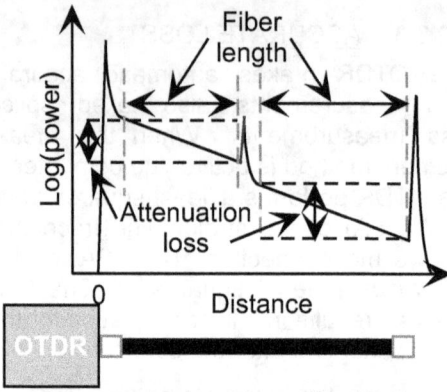

Figure 4-20: Correct Attenuation Rate Measurement With Non-Uniformity

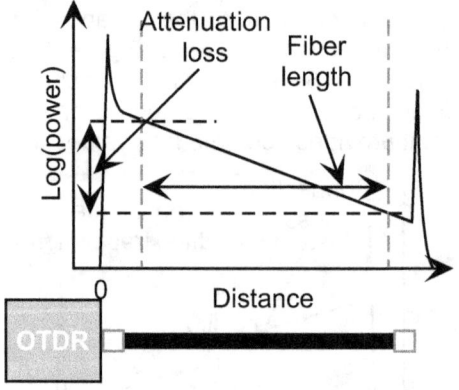

Figure 4-21: Reduced Cursor Separation

To understand this, remember that a properly designed, manufactured, and installed cable segment will have a uniform trace. If the trace is uniform and high, the problem is defective cable. If there is defective installation, the location of the defect is at the non-uniform loss (Figure 4-18 and Figure 2-15).

4.2.5 CURSOR IN PEAK

Upon review of all the cursor placement figures in the last sections, one important rule becomes evident:

> ➤ Cursors are never placed in a reflection. *There is no valid measurement with cursors in a reflection.*

4.3 DIRECTIONAL DIFFERENCES

Usually, insertion loss measurements are different in opposite directions. In 3.3, you learned that gainers result from differences in the fibers at a connection. From these two facts, it is reasonable to expect that OTDR loss measurements in opposite directions will result in different values. Such is the case.

4.4 DUAL TRACES

4.4.1 DUAL WAVELENGTH TRACES

Some OTDR software allows display of two traces of the same fiber. With this feature, an operator can compare the traces for two wavelengths. This comparison enables immediate identification of fiber under stress.

4.4.2 BI-DIRECTIONAL TRACES

Some OTDR software allows display of bi-directional traces of the same fiber. With this feature, an operator can compare the losses of splices visually. In addition, such a feature enables automatic software calculation of average splice loss.

4.5 ADJUSTING MEASUREMENTS

As software is not always perfect, the automatic table values can be incorrect. Some OTDRs allow correction of such inaccurate measurements by adjusting cursor locations.

4.6 TRACE COMPARISON

Occasionally, the operator will benefit from comparing traces on fibers in the same cable. Using a multiple trace display option, similar to dual wavelength trace display option, the operator can determine whether an event has the same characteristic on multiple fibers. With such a determination, the operator can differentiate between a high loss splice and a routing problem. A splice problem consistent in multiple fibers is unusual. A routing problem consistent in multiple fibers is not unusual.

4.7 FTTH/PON LINK TRACES

Thus far, all of the traces presented have point-to-point topologies. The FTTH/PON link differs from this topology, in that the FTTH/PON link is point to multi-point. This difference complicates in interpretation of the OTDR trace in one direction.

From the central office (CO) to the subscriber optical line terminal (OLT), the 'downstream' direction, the link is point to multipoint (Figure 4-22). In the reverse direction, the 'upstream' direction, the link is point to point.

In the downstream direction, light travels through the fiber, through a splitter and into multiple output fibers (Figure 4-22). The backscattered power travels from each fiber, through the splitter back to the OTDR. This travel path produces a trace like that in Figure 4-23. Note that the first section of the backscatter after the splitter includes backscatter from two fibers, while that from the second section includes backscatter from one fiber.

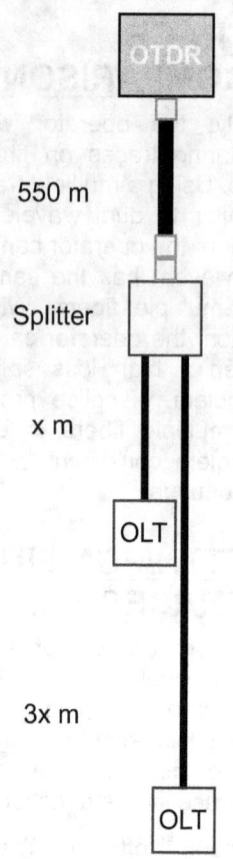

Figure 4-22: FTTH/PON Link With 1x2 Splitter

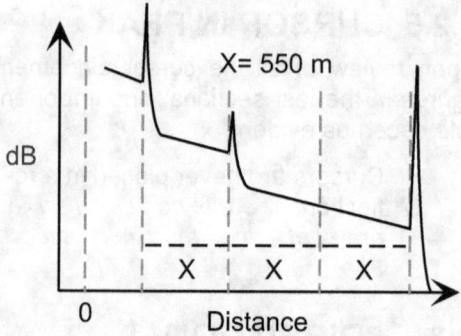

Figure 4-23: Downstream Trace Through 1x2 Splitter

4.8 KEY CONCEPTS

1. Set up requires seven types of information.
2. The launch cable core diameter must match that in the cable under test.
3. The launch cable must have low loss connectors.
4. The launch cable must have low reflectance connectors.
5. Use a launch cable longer than the dead zone.
6. Make all measurements with scales expanded.
7. Cursors are always placed in the backscatter.
8. Cursors are never placed in the reflections.
9. A length measurement includes a peak or a drop off at the beginning and excludes a peak or drop off at the end.
10. An estimated connection loss measurement of reflective or non-reflective connections requires two cursors placed in the backscatter on both sides of the reflectance or drop off, but as close as possible to the reflectance or drop off without being in the reflectance or drop off.
11. For far end connector measurement loss, the operator moves the OTDR to the far end.
12. Alternatively, for far end connector measurement loss, the operator installs a second 'launch' cable on the far end.
13. An attenuation rate measurement requires placement of two cursors near the ends of the same

The trace in Figure 4-23 results from a simple configuration of a 1 x 2 splitter. In an FTTH/PON, the link could be a single splitter with up to 64 output fibers or multiple splitters with a total split of 64. The features of such a trace are the sum of reflections and backscatter from multiple link components. The interpretation of such traces can be difficult and complex. This difficulty makes troubleshooting a subscriber problem from the CO more difficult than troubleshooting a point-to-point link.

In the upstream direction, the features of the trace are created by components of a single link. As such, it is possible to measure attenuation rates, connection and splitter losses and, in general, troubleshoot from the subscriber OLT to the CO.

straight-line segment without enclosing any features.

14. If the attenuation rate is high and the backscatter trace is straight, i.e., uniform, the problem is defective cable.

15. If the attenuation rate is high and the backscatter trace is not straight, the problem is defective installation.

16. Simultaneous display of traces at both wavelengths can reveal fiber under stress.

17. Simultaneous display of traces from both directions can allow operator to make visual review of splice loss in both directions.

18. Simultaneous display of traces from multiple fibers can reveal the difference between high loss splices and routing problems.

5 INTERPRETATION

Chapter Objective: you learn how to determine a maximum acceptance values and to interpret measured loss with the acceptance values. In addition, you see
the effects of improper cursor placement on a trace.

5.1 CERTIFICATION STRATEGY

Acceptance values are the maximum loss values that will be accepted. To interpret the measured loss values, the operator must have a strategy for determining ac-ceptance values. The operator chooses from at least four strategies.

➢ Maximum loss values

➢ Typical Loss values

➢ Midpoint values

➢ Statistical values

The data standards allow use of the first strategy, i.e., maximum loss values, as acceptance values. However, most prod-ucts test closer to their typical loss than to their maximum loss. As a general rule, components have loss closer to the maximum value than to the typical value have been installed incorrectly. Incorrectly installed components have reduced relia-bility. Thus, use of the maximum loss val-ues as acceptance values can result in acceptance of conditions of reduced reli-ability (PFOI, Ch. 19).

Instead of maximum loss values, the op-erator could use the typical loss values. This author has had conversations with professional designers and installers who have used this strategy.

Use of the typical loss values as maxi-mum acceptance values can result in in-creased cost with no benefit. Properly installed cable, connectors, and splices have a distribution of loss, with some val-ues below the typical and some, above. Use of typical loss as an acceptance value will result in rejection of properly installed products with loss slightly above the typical value. There is no benefit with this strategy.

The third strategy is use of acceptance values half way between the maximum and typical values. This strategy, a 'mid-point' strategy, is based on the assump-tion that properly installed components will test closer to the typical value than the maximum value most of the time. This author recommends this strategy. With the mid-point strategy, acceptance values are those in Table 5-1 and Table 5-2.

The fourth strategy requires statistical information about the loss of the products. With statistical information, the operator can calculate a statistically valid upper limit on the loss. At the time of this writing, this strategy has little use.

Wavelength, nm	Core Diam-eter, μ	Attenuation Rate, dB/km
850	62.5	3.15-3.25
850	50	3.0-3.1
1300	62.5	1.1
1300	50	1.1
1310[3]	8.2	0.4-0.425
1310[4]	8.2	No data
1550[2]	8.2	0.35
1550[2]	8.2	No data

Table 5-1: Attenuation Rate Acceptance Values

Installation Method	Ferrule mm	dB /Pr
Polish	2.5	0.525
Polish	1.25	0.475
Cleave and crimp		0.588
Splice		0.125

Table 5-2: Connection Acceptance Values

5.2 EXAMPLE, END 1

Figure 5-1 presents a link of two seg-ments, each of 100-110 m, connected by a 1 m patch cable. The launch cable is 36.3 m long.

---

[3] Loose tube cable

[4] Tight tube cable

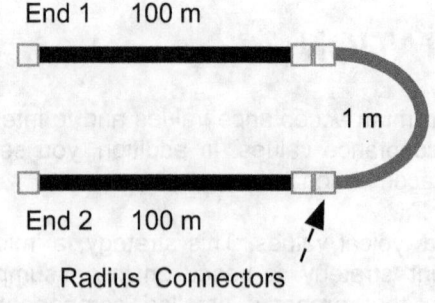

End 1   100 m

1 m

End 2   100 m

Radius Connectors

Figure 5-1: Link Map

Figure 5-2 is a trace of this link from end 1. This trace is different from what would be expected from the map. Two 'events', at approximately 70 m and 175 m, are multiple, or ghost, reflections. In 5.3, we verify that these unexpected trace features are, indeed, ghosts.

Note two aspects of the traces from the Tektronix 3031 mini-OTDR. The first is that the measurements are all indicated in the box at the center and top of the screen. The second is that a short horizontal line indicates the location at which each cursor crosses the trace. This short line simplifies correct cursor placement.

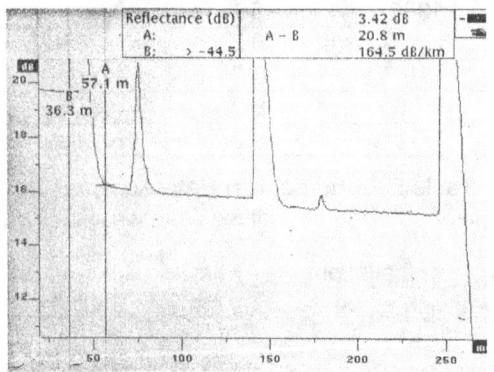

Figure 5-2: Trace From End 1

5.2.1 CONNECTION LOSS

Figure 5-2 presents correct cursor placement for loss measurement of the end 1 connector. The loss is 3.42 dB. This value exceeds the acceptance value (Table 5-2).

Figure 5-3 presents correct cursor placement for loss measurement of the mid span connectors. The loss is 0.32 dB. As the separation between the cursors, 20.8 m, is greater than the length of the patch cord, 1m, this reflection includes two

pairs. With an acceptance value of 1.05 dB, we accept these two connector pairs.

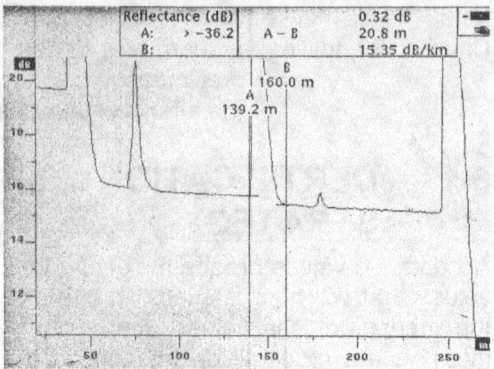

Figure 5-3: Mid-Span Connector Loss

The connector at end 2 cannot be measured from the near end. However, it can from the far end. (5.3).

5.2.2 ATTENUATION RATE

Figure 5-4 presents correct cursor placement for attenuation rate measurement of segment 1. The value is 3.318 dB/km. Note that the cursors do not include the entire segment. By the rules of cursor placement, the ghost reflection at approximately 70 m must be excluded.

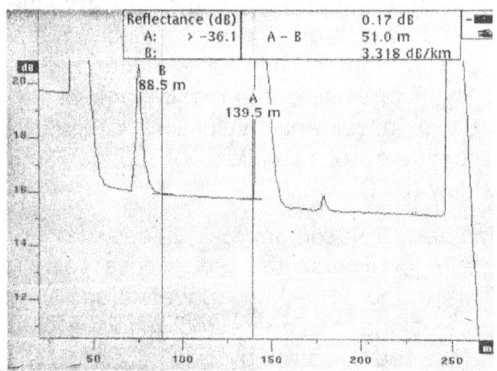

Figure 5-4: Segment 1 Attenuation Rate

Figure 5-5 presents correct cursor placement for attenuation rate measurement of segment 2. The value is 2.751 dB/km. Note that the cursors do not include the entire segment. By the rules of cursor placement, the ghost reflection at approximately 175 m must be excluded.

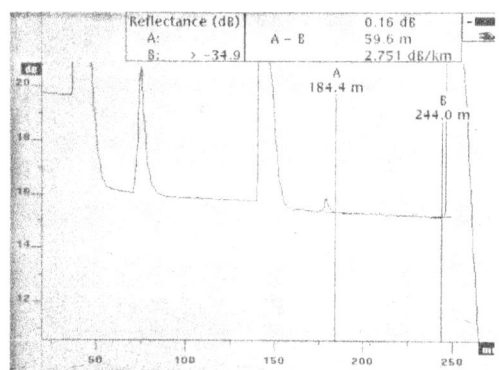

Figure 5-5: Segment 2 Attenuation Rate

An acceptable attenuation rate has two requirements:

➤ Uniform loss

➤ Loss less than the acceptance value

With the exceptions of the multiple reflections, the traces of both segments are straight, indicating uniform attenuation.

With an acceptance value of 3.25 dB/km, we accept segment 2 (2.751 dB/km), but not segment 1 (3.318 dB/km). Since the trace for segment 1 is uniform, the most likely interpretation is that this segment has defective cable. Error in installation would result in a non-uniform trace. We re-examine this conclusion in 5.3.

5.2.3 LENGTHS

Figure 5-6 presents correct cursor placement for length measurement of segment 1. The length is 103.2 m.

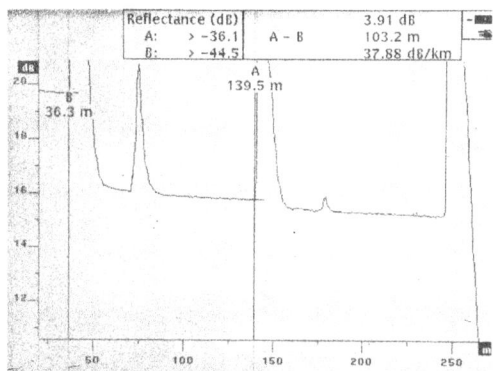

Figure 5-6: Segment 1 Length

Figure 5-7 presents correct cursor placement for length measurement of segment 2. The length is 106.0 m.

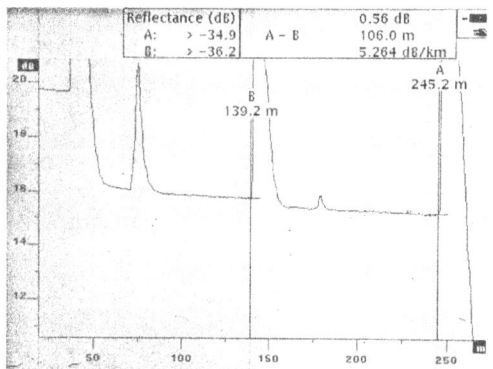

Figure 5-7: Segment 2 Length

Being between 100 and 110 m, both lengths meet the requirements of the map.

5.3   EXAMPLE, END 2

Figure 5-8 presents correct cursor placement for loss of the end 2 connector. The value is 0.18 dB. This value is acceptable, since it is less than the acceptance value (0.525 dB/pr).

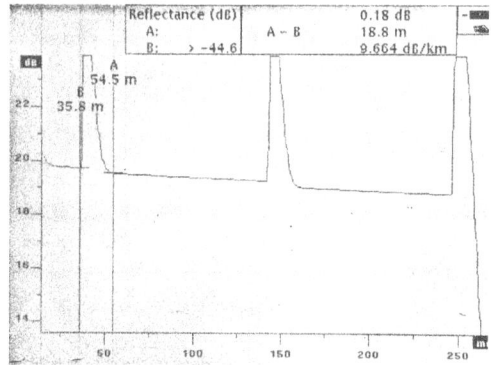

Figure 5-8: Connector Loss End 2

This figure is not consistent with all figures from end 1 (Figure 5-2 to Figure 5-7). In Figure 5-2, two apparent ghosts appear. In Figure 5-8, they do not. With this comparison, we have proven that the reflections at approximately 70 and 175 m from end 1 are not real; they are multiple reflections. Were they real fiber ends, they would create reflections in both directions.

These two figures demonstrate one of the characteristics of multiple reflections: the closer to the noise floor the trace is, the more likely a ghost can appear. In this case, the trace from end 1 had a high loss connector at end 1 (3.42 dB). The ghost appeared. The trace from end 2 has a low loss connector (0.18 dB). No ghost appeared.

In addition, the presence of the ghosts results in inaccurate attenuation rate measurements. Figure 5-9 indicates a segment 1 attenuation rate of 2.905 dB/km. This value is less than the acceptance value.

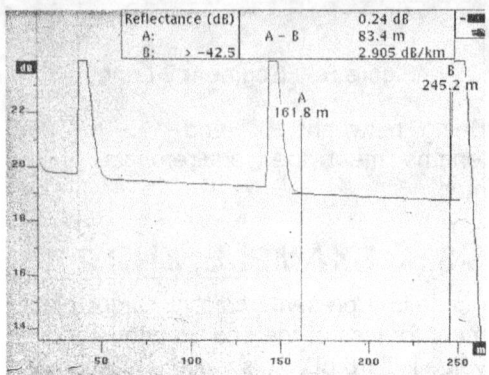

Figure 5-9: Attenuation Rate, Segment 1 From End 2

This value is lower than the value measured from end 1, 3.318 dB/km (Figure 5-4). Thus, the conclusion we made from the attenuation rate measurement from end 1 was incorrect (5.2.2).

Note that the ghost did not significantly change the attenuation rate of segment 2. From end 1, the value is 2.751 dB/km. From end 2 [Figure 5-10], the value is 3.056 dB/km.

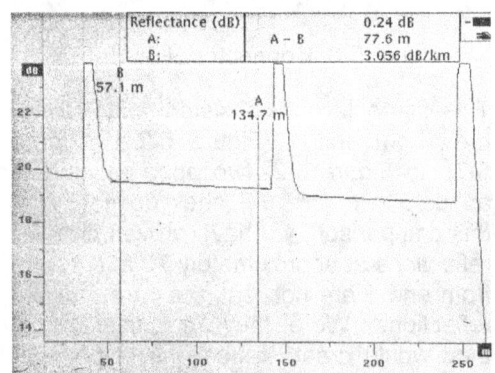

Figure 5-10: Segment 2 Attenuation Rate, End 2

5.4 INCORRECT CURSOR PLACEMENT

In this section, we examine the effects of incorrect cursor placement.

5.4.1 LENGTH

Figure 5-11 presents incorrect cursor placement for length measurement. The value indicated, 104.0 m, is greater than that for correct placement, 103.2 m (Figure 5-6).

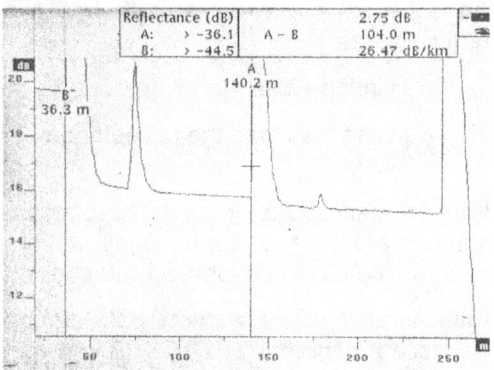

Figure 5-11: Incorrect Cursor Placement For Length

5.4.2 ATTENUATION RATE

Figure 5-12 presents incorrect cursor placement for attenuation rate measurement. The value indicated, 5.484 dB/km, is greater than that for correct placement (Figure 5-4).

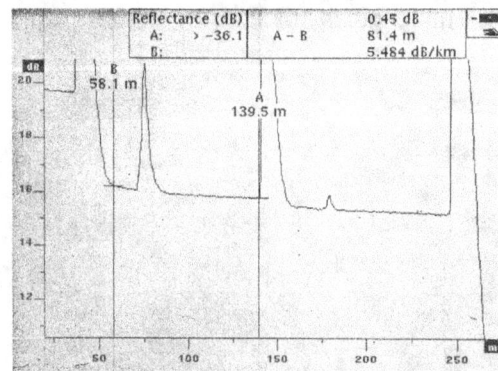

Figure 5-12: Incorrect Cursor Placement For Attenuation Rate

5.4.3 CONNECTOR LOSS

Figure 5-13 presents incorrect cursor placement for connector loss measurement. The value indicated, 0.37 dB, is greater than that for correct placement, 0.32 dB (Figure 5-3).

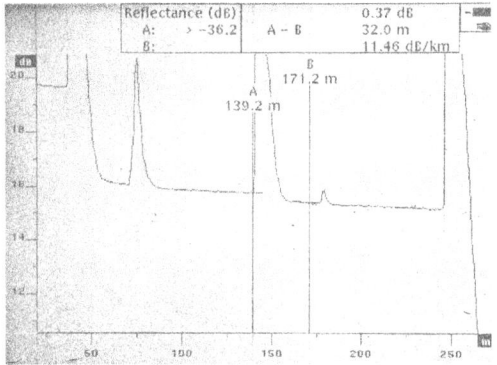

Figure 5-13: Incorrect Cursor Placement For Connector Loss

5.5   LENGTH CONSISTENCY

As would be expected, length measurements of the same link in the opposite directions will not always be the same.

Appendices 8.7 and 8.8 provide examples of this fact. Appendix 8.7 presents length measurements from opposite directions. While some are exactly the same, many are not.

Appendix 8.8 presents distance to splices in opposite directions. Note that the 'real distance' to events was not the same in both directions. Comparison of the total length in both directions, 164.295 and 164.227 kft, indicates a length difference of 0.068 kft.

5.6   KEY CONCEPTS

1. The OTDR operator must have a certification strategy to determine acceptance values.
2. The mid-point strategy provides maximum acceptance values closer to the typical than to the maximum.
3. Multiple, or ghost, reflections can cause inaccurate loss measurements.
4. Ghost reflections occur after high loss connectors but may not occur after low loss connectors.

6   TESTING PRACTICALITIES

Chapter Objective: you learn how to make and interpret trace data, both qualitatively and quantitatively.

6.1   ACQUIRE LINK DATA

For all cables to be tested, acquire the following information:

➢ Link map

➢ Core diameter

➢ Wavelength(s) of operation

➢ Attenuation rates

➢ Connector loss

➢ Splice loss

6.2   SET UP

Note: each OTDR has different set up instructions. The following steps are generic and may not be complete for a specific OTDR.

Enter ore select the following values into the OTDR menu(s):

➢ Wavelength

➢ Pulse width

➢ Maximum cable length to be tested

➢ Either the maximum time allowed for the test or the number of pulses to be analyzed

➢ Index of refraction (8.3)

➢ Backscatter coefficient (8.2)

➢ Measurement thresholds (vary from OTDR to OTDR)

➢ Acceptance values for automatic trace analysis for: attenuation rates, splice loss, connector loss

This author uses an event loss threshold of 0.05 dB. This value results in no table entries for non-reflective events, such as fusion splices, if the loss is less than this value.

Use of a value less than 0.05 dB has two consequences:

First, noise in the signal may create automatic table events that are not real. Second, there will be increased work to average splice loss, as shown in Appendix 8.8.

6.3   LAUNCH CABLE

Clean OTDR port and both ends of launch cable. Periodically, inspect both connectors on launch cable with a microscope.

Turn on or off automatic trace analysis function.

Run trace. If the OTDR has an internal fiber, record the OTDR port loss for reference. Record the launch cable length.

Compare current port loss and launch cable length to previous values. If there is a significant increase in loss or change in length, resolve changes before testing.

Port loss should be approximately 0.3 dB. Cable length should not change more than 1 or 2 m. If port loss becomes high, the maximum length of cable can become reduced. In addition, ghost reflections are increasingly likely to appear.

A significant increase for port loss is 0.1-0.2 dB. An increase greater than 0.2 dB may indicate dirt on OTDR port or launch cable. Clean both ends and retest launch cable.

If cleaning does not correct increase in loss, test launch cable from opposite end. If increase remains, test back up launch cable. If increase disappears, launch cable requires maintenance.

If increase cannot be reduced, replace launch cable. A high loss connector on launch cable will bias near end connector loss measurement higher than the real value.

Measure and record dead zone, for reference. The dead zone will vary with wavelength and pulse width.

Review map to determine locations at which dead zone contains more than one component.

> Example: a patch panel with a 2 m patch cord between two long segment lengths

Record those locations for comparison to values in table generated automatically.

6.4 QUALITATIVE EVALUATIONS

Connect cable.

Run trace. Make a qualitative evaluation of all features on trace. Specifically, look for:

> Low noise

> Uniformity of attenuation rate

> Small dead zone with clear far end reflectance

> Height of reflective events

> No overshooting

> Multiple reflections

If trace has low noise, each cable segment will have a straight line trace. If trace appears noisy, change settings of pulse width, number of pulses, or time for trace. Rerun trace. Make additional changes until trace is noise free.

Compare dead zone at end of launch cable to those along the cable. Dead zones should be approximately the same width. If the dead zone at end of launch cable is much larger than those along the cable, there may be a high reflectance connector. Clean both connectors and retest.

The height of reflection connections reflects the reflectance of such connections. Comparison of the heights of multiple fibers will indicate whether the launch cable connector or the connector on the cable under test has a high reflectance. Cleaning of both connectors can eliminate high reflectance.

Overshooting may indicate high reflectance connectors. Clean both and retest. If overshooting persists, increase pulse width until overshooting does not appear.

Multiple reflections may indicate that the connector on the launch cable is high reflectance. Cleaning of the connectors or replacement of the launch cable may eliminate these reflections.

6.5 QUANTITATIVE EVALUATIONS

The OTDR operator can perform quantitative evaluations automatically or manually. Automatic evaluations require software that functions properly. The operator is well advised to verify proper operation prior to acceptance of automatically generated values.

6.5.1 AUTOMATIC MEASUREMENTS

Compare table to trace. All events on trace should appear in table and vice versa.

Compare the table indication of acceptance to the known acceptance values. Since a single acceptance value may not apply to all events on a trace, the operator applies separate values to events that represent multiple connections in a singe reflection or drop.

The OTDR operator should compare table values to manually measured values. This comparison should be done prior to acceptance of table values. In addition, the operator should make spot checks of table values against manually measured values.

When making spot checks of connection loss, the operator should remember that manual loss measurements can include attenuation in the fiber between cursor positions. Such loss increases the manual loss above table values.

6.5.2 MANUAL MEASUREMENTS

Make manual measurements according to the cursor placement rules in Section 4.2.

Loss of near end connector includes loss of launch cable connector. If near end connector loss is acceptable, continue using launch cable.

Near end connector loss should be 0.2 to 0.4 dB/pair. For connectors with the 1.25 mm ferrule, such as LC, MU, and LX.5, typical loss is 0.2 dB. For connectors with the 2.5 mm ferrule, such as ST-™ compatible, SC, FC, and OptiJack, typical loss is 0.3 dB. If the connector was installed with a cleave and crimp installation method, the typical loss is 0.4 dB.

6.6 DUAL WAVELENGTH TESTS

Whether required or not, testing at both wavelengths is advisable. Increased attenuation rate at the long wavelength can indicate installation errors. In addition, increased loss at the long wavelength can indicate the nature of the corrective action for splices. Increased loss at the long wavelength can indicate a routing problem. The same high loss at both wavelengths can indicate the need to re-splice the fibers.

6.7 BI-DIRECTIONAL TESTS

Whether required or not, testing both directions is necessary in order to accept splices and connectors. To accept a splice with a gain, the operator must test the splice in both directions. To accept a high loss splice, the operator must test the splice in both directions and average the losses.

6.8 SPLICE LOSS AVERAGING

Splice loss averaging is simple...if you have an accurate map. With such a map, the operator can calculate the distance of each splice from both ends of the link. Then, he compares the locations indicated in the OTDR traces or the tables to the calculated distance. In an ideal world, the calculated distance and the map distance are the same.

In the real world, there may be differences. Small differences are due to the limitations of the OTDR. With some experience, the operator will develop an estimate of the typical measurement error for his OTDR. If the difference between the map distance and the OTDR distance is close this this estimate, the operator will assume that the calculated and OTDR locations are the same. This author's experience is that OTDR length measurements can differ by 100-200 feet. In addition, this difference increases with increasing cable length.

If the difference is greater than this range, differences may indicate different splices, the losses of which should not be averaged. In this case, the second splice value is 0 dB.

This difference may indicate error in the map. Such error is not unusual.

Such difference may result from the difference between cable and fiber lengths. One method to reduce errors from this cause is measurement of the distance from the nearest known splice or patch panel. As the distance between events becomes smaller, the distance error due to fiber and cable length difference becomes smaller.

7 REVIEW QUESTIONS

7.1 OTDR PRINCIPLES

1. At what wavelengths does an optical link operate?

2. What is the name of the cause by which power is lost as it travels through a fiber?

3. Is there technical support for use of an OTDR to measure the loss that a transmitter-receiver pair will experience?

4. How many reasons are there for your answer to the previous question?

5. What are the reasons for your answer to the previous question?

6. What does the OTDR provide?

7. Can we simultaneously transmit the same wavelength in opposite directions on the same fiber?

8. Justify your answer to the previous question.

9. What is the name of the power from the core returns to the OTDR?

10. What does IR mean?

11. What are the units of measure of the IR?

12. What is the range of values of the IR?

13. How does the OTDR operator use the IR?

14. AN OTDR operator sends a splicer to repair a splice at an OTDR distance of 5600 m. The splicer tells the operator that there is no splice enclosure at that location. What fact did the operator forget?

15. What are the two sources of optical power that return to the OTDR?

16. Is an insertion loss test sufficient to ensure reliable link operation?

17. Justify your answer to the previous question.

18. What is the vertical axis of an OTDR trace?

19. What are the units of the vertical axis?

20. What is the horizontal axis of an OTDR trace? Be careful.

21. Does the OTDR measure the units displayed on the horizontal axis?

22. What five measurements might you make with an OTDR?

23. What measurement might you not be able to make with an OTDR?

24. _____ are known as peaks and spikes.

25. Of the three terms in the previous question, which is the correct technical term?

26. List five causes of peaks from high peak to low or no peak.

27. A _____ zone is also known as a _____ zone.

28. You observe an OTDR operator making loss measurements of the zones in the previous question. You observe that he has no map. What do you suspect about these loss measurements?

29. Justify your answer to the previous question.

30. Select the best answer. The reflection in the circle of the figure below could have:

    ___ One connector pair
    ___ Two connector pairs
    ___ Any number of connector pairs

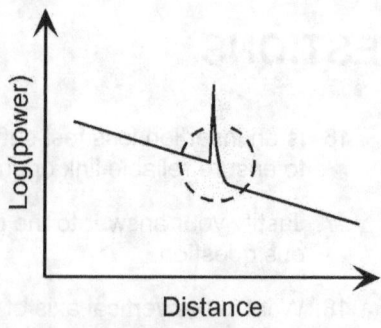

Distance

31. What fact must be true to answer the previous question?

32. The attenuation rate in segment A is: ___ higher / ___lower than that in Segment B.

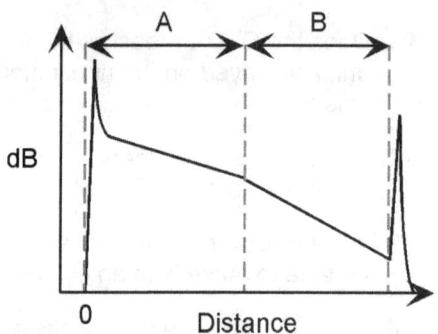

7.2   BASIC TRACES

1. What five-word answer results in a reflectance?

2. How many basic traces are there?

3. What is each of the basic traces in the previous question?

> For Questions 4-11, you will need only three different words.

4. Radius connectors _____ create reflectance.

5. Singlemode mechanical splices connectors _____ create reflectance.

6. Multimode mechanical splices connectors _____ create reflectance.

7. Multimode fusion splices connectors _____ create reflectance.

8. Singlemode fusion splices connectors _____ create reflectance.

9. APC connectors _____ create reflectance.

10. Broken fibers connectors _____ create reflectance.

11. A violation of a cable performance parameter that does not result in a broken fiber connectors _____ creates reflectance.

12. A link map indicates two segments. The trace indicates three segments. How do you explain this difference?

13. What are the two characteristics of a 0 dB fusion splice?

> For Questions 14-20, multiple answers are possible. In addition to indicating whether or not a type of feature can be present, provide the reason that it can, or cannot, be present.
>
> In the context of these questions, a bad cleave is a high angle cleave that reflects power outside the critical angle.

14. The feature(s) in the circle below could be:

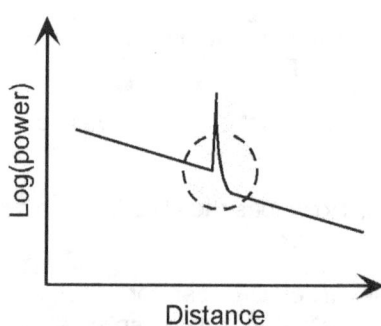

Distance

___ Mechanical splice, multimode
___ Mechanical splice, singlemode
___ Fusion splice, multimode
___ Fusion splice, singlemode
___ Fusion splice with gas bubble
___ APC connectors
___ Radius connectors

___ Single segment
___ Two segments
___ Bend radius violation
___ Broken fiber in dead zone

15. The feature(s) in the circle below could be:

___ Mechanical splice, multimode
___ Mechanical splice, singlemode
___ Fusion splice, multimode
___ Fusion splice, singlemode
___ Fusion splice with gas bubble
___ APC connectors
___ Radius connectors
___ Single segment
___ Two segments
___ Bend radius violation

16. The feature(s) in the circle below could be:

___ Mechanical splice, multimode
___ Mechanical splice, singlemode
___ Fusion splice, multimode
___ Fusion splice, singlemode
___ Fusion splice with gas bubble
___ APC connectors
___ Radius connectors
___ Single segment
___ Two segments
___ Bend radius violation
___ Broken fiber in dead zone

17. The feature(s) in the circle below could be:

___ Mechanical splice, multimode
___ Mechanical splice, singlemode
___ Fusion splice, multimode
___ Fusion splice, singlemode
___ Fusion splice with gas bubble
___ APC connectors
___ Radius connectors
___ Single segment
___ Two segments
___ Bend radius violation
___ Broken fiber at OTDR
___ Broken fiber in dead zone

18. The feature(s) in the circle below could be:

___ Mechanical splice, multimode
___ Mechanical splice, singlemode
___ Fusion splice, multimode
___ Fusion splice, singlemode
___ Fusion splice with gas bubble
___ APC connectors
___ Radius connectors
___ Single segment
___ Two segments
___ Mechanical splice on cable end
___ Bend radius violation
___ Broken fiber in dead zone

19. The feature(s) in the circle below could be:

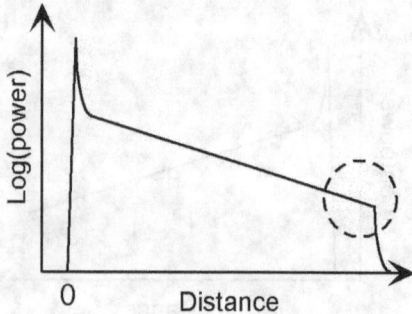

___ Mechanical splice, multimode
___ Mechanical splice, singlemode
___ Fusion splice, multimode
___ Fusion splice, singlemode
___ Fusion splice with gas bubble
___ APC connectors
___ Radius connectors
___ Single segment
___ Two segments
___ APC connector on cable end
___ Bad cleave on cable end
___ Mechanical splice on cable end
___ Bend radius violation
___ Broken fiber in dead zone

20. The feature(s) in the circle below could be:

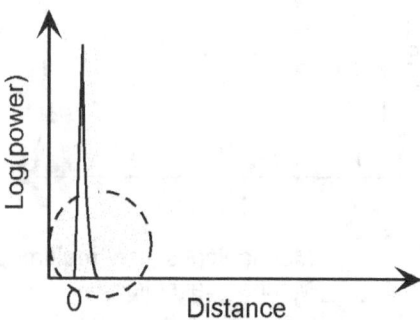

Hint: there is a simple, three or four word, answer to this question.

___ Mechanical splice, multimode
___ Mechanical splice, singlemode
___ Fusion splice, multimode
___ Fusion splice, singlemode
___ Fusion splice with gas bubble
___ APC connectors
___ Radius connectors
___ Single segment

___ Two segments
___ APC connector on cable end
___ Bad cleave on cable end
___ Mechanical splice on cable end
___ Bend radius violation
___ Broken fiber at OTDR
___ Broken fiber in dead zone

7.3 UNUSUAL TRACES

1. How many potential causes are there of a trace end without a reflection?

2. What are the causes in the previous question?

3. Can gainers occur at connectors?

4. Justify your answer to the previous question.

5. After the test at both wavelengths of a link, a splice exhibits high loss. Your splice operator sees the values and tells you he will go to correct this condition. What should you say to him?

6. Justify your answer to the previous question.

7. The OTDR operator makes a trace at one wavelength. The splicer sees the high value and starts to leave to remake the splice? What is your best technical recommendation?

8. A non-linear trace indicates _____ on the fiber.

9. When properly installed, the attenuation rate at a long wavelength will be _____ than that at a short wavelength.

10. When improperly installed, the connection loss at a long wavelength will be _____ than that at a short wavelength.

11. How many types of measurements can become inaccurate due to high reflectance?

12. What are the types in the previous question?

13. What is the key trace characteristic that indicates overshooting?

14. What is the technical reason that overshooting is theoretically impossible?

15. How many potential solutions to overshooting?

16. What are the potential solutions to overshooting?

7.4 TEST AND MEASUREMENTS

1. The OTDR operator must input ____ settings.

2. What are the settings in the previous question?

3. What can happen to average splice loss if the loss threshold is 0.05 dB?

4. The launch cable has ____ functions.

5. What are the functions in the previous question?

6. What is the absolute minimum length of a launch cable?

7. According to the text, what is a reasonable length for a launch cable?

8. What problem can a launch cable create?

9. How do you avoid the problem in the previous question.

10. The launch cable has ____ requirements.

11. What are the requirements in the previous question?

12. What one step should be taken during all measurements?

13. What are the two features on a trace that indicate a segment end?

14. State the rules of cursor placement for attenuation rate measurements.

15. State the rules of cursor placement for estimated connection loss measurements.

16. State the rules of cursor placement for accurate connection loss measurements.

17. State the rules of cursor placement for segment length measurements.

18. Is the estimated connection loss greater than or less than the accurate connection loss?

19. Explain your answer to the previous question.

20. What measurement is made with cursor(s) placed in a reflectance?

21. Justify your answer to the previous question.

22. You can measure the loss of a connector on the far end of a cable can in ____ ways.

23. What are the ways in the previous question?

For each of the traces in Questions 24-31, identify the type of measurement.

Use the following codes:

X= invalid cursor placement
AR= attenuation rate
AC= accurate connection loss
EC= estimated connection loss
FS= first segment length measurement
SS= subsequent segment length

24. The type of measurement is _____

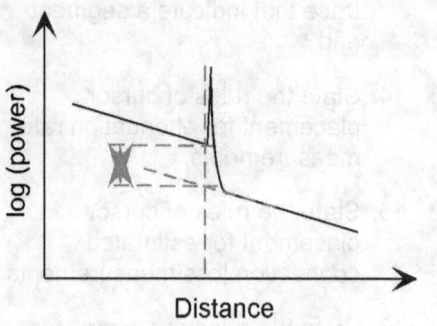

25. The type of measurement is _____

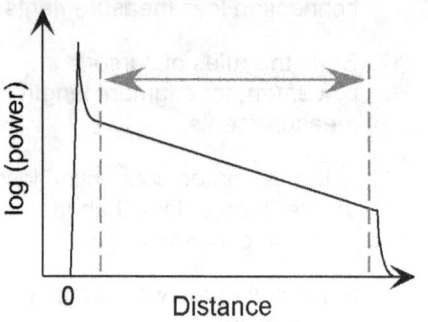

26. The type of measurement is _____

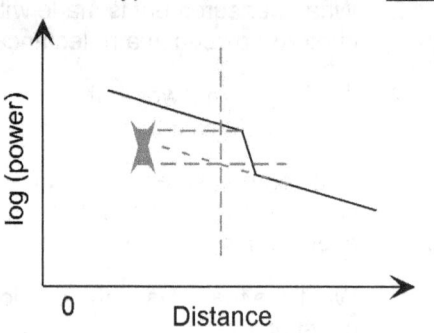

27. The type of measurement is _____

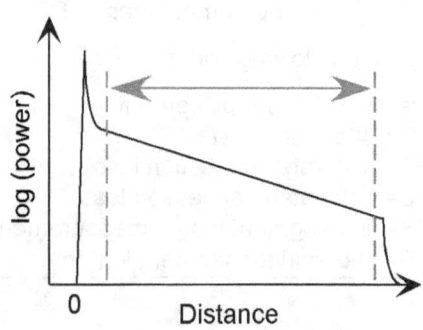

28. The type of measurement is _____

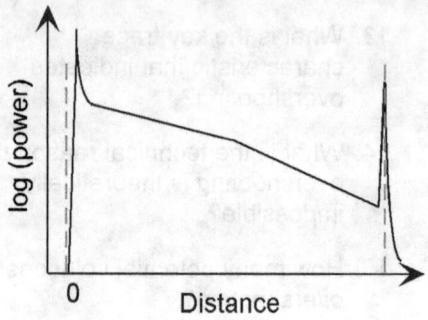

29. The type of measurement is _____

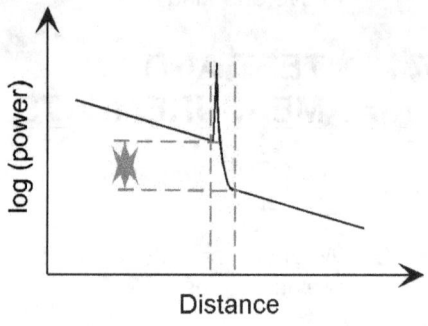

30. The type of measurement is _____

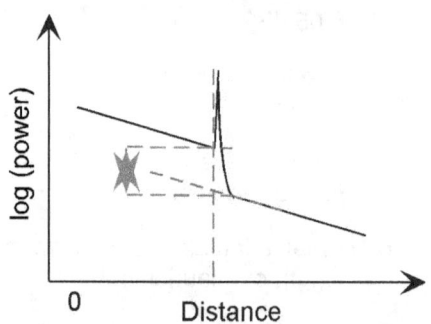

31. The type of measurement is _____

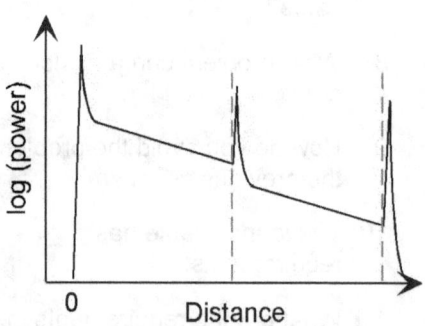

For questions 32 through 39, draw the trace you would expect. Assume a launch cable of 50 m. Unless otherwise indicated, assume that all connections have observable loss. Assume that the dead zone is 20 m. Use the key in Figure 1-12.

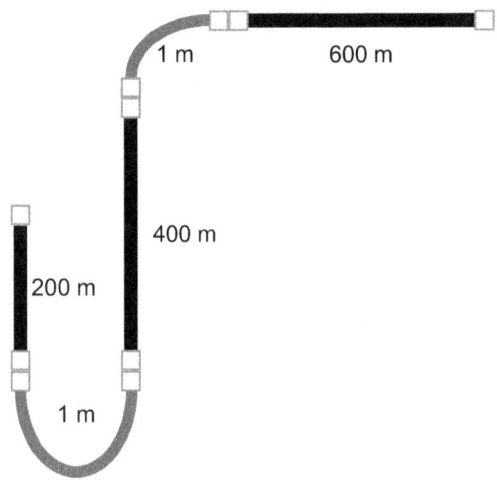

Figure 7-1: Map To Trace 1

32. Draw the trace from the 200 m end [Figure 7-1] Use the figure below.

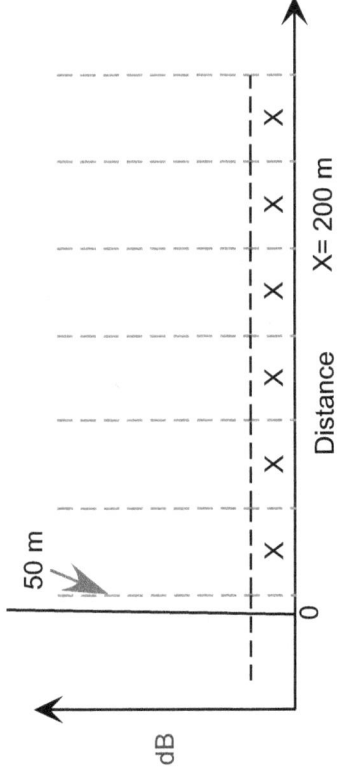

Figure 7-2: Answer Form For Question 32

33. Draw the trace from the 600 m end [Figure 7-1] Use the figure below.

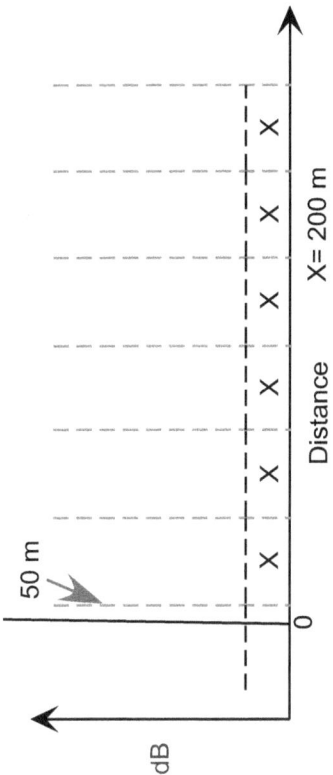

Figure 7-3: Answer Form For Question 33

34. Draw the trace in Question 32 with a ghost reflection from the launch cable [Figure 7-1] Use the figure below.

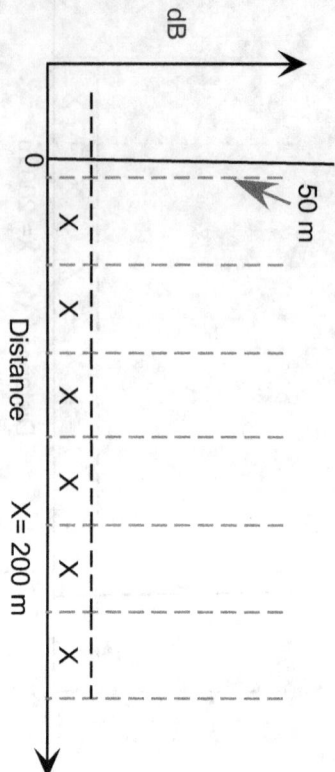

Figure 7-4: Answer Form For Question 34

35. Draw the trace in Question 33 with a ghost reflection from the launch cable [Figure 7-1] Use the figure below.

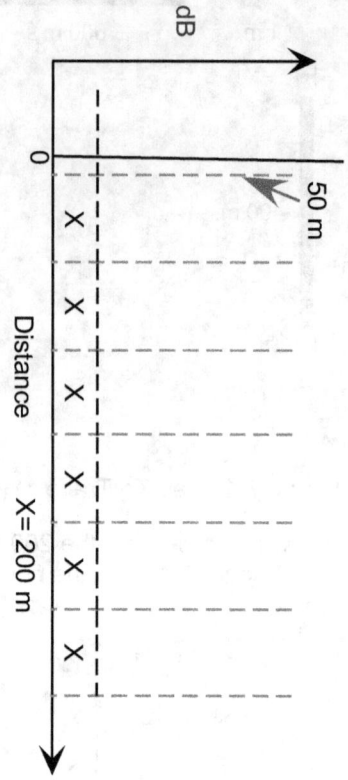

Figure 7-5: Answer Form For Question 35

When tested from end 1, the gainer near end 1 is a gainer. When tested from end 2, the gainer near end 2 is a gainer. The launch cable is 50 m with a dead zone of 20 m.

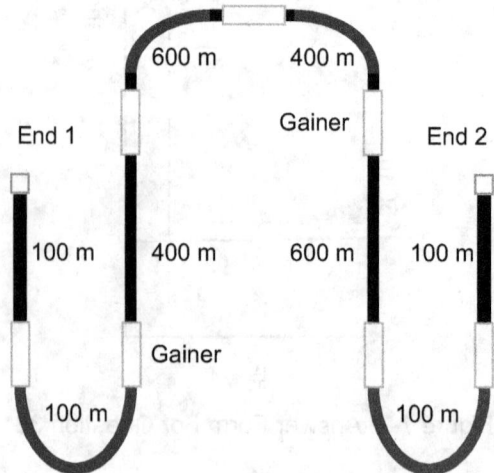

Figure 7-6: Map To Trace 2

36. Draw the trace from End 1 [Figure 7-6]. Use the figure below.

37. Draw the trace from the End 2 [Figure 7-6]. Use the figure below.

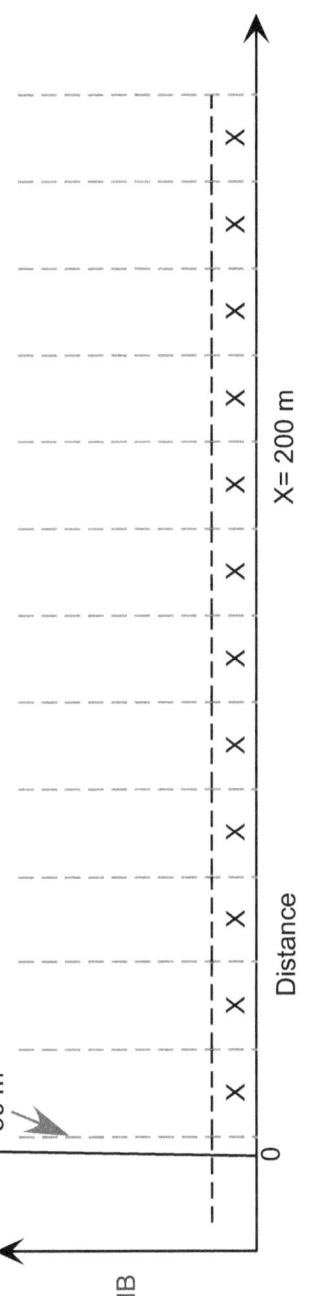

Figure 7-7: Answer Form For Question 36

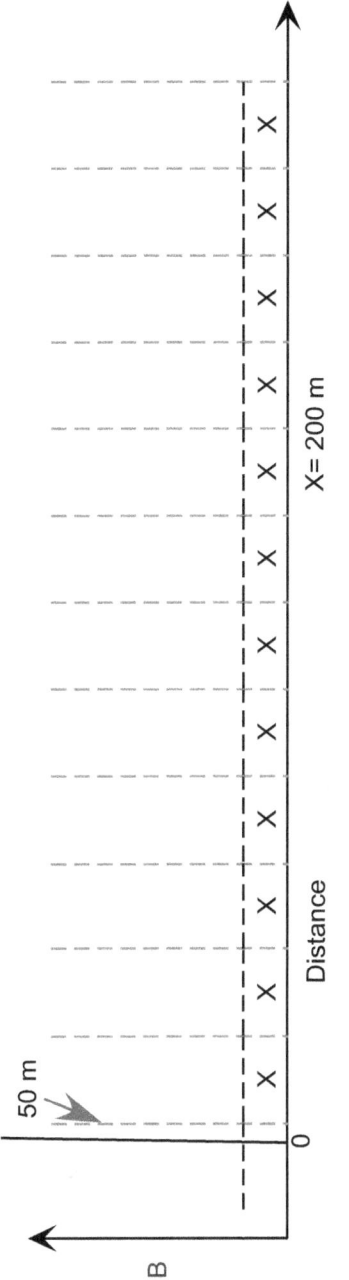

Figure 7-8: Answer Form For Question 37

7.5 INTERPRETATION

1. Figure 5-1 is the link map. Table 5-1 and Table 5-2 have the acceptance values. Figure 7-9 to Figure 7-22 to contain traces from end 1. Figure 7-23 contains the trace from end 2.

For each of these traces, indicate the following about cursor placement:

Figure	Valid?	Type
7-9		
7-10		
7-11		
7-12		
7-13		
7-14		
7-15		
7-16		
7-17		
7-18		
7-19		
7-20		
7-21		
7-22		
7-23		

Table 7-1: Cursor Validity And Measurement Type

2. Then use Figures to certify the link. Use table 7-2 to indicate your results.

	Fig. #	Value	Accept?
Connector End 1			
Connector End 2			
Connector, Middle			
Length, Segment 1			
Length, Segment 2			
Attenuation Rate, Segment 1			
Attenuation Rate, Segment 2			

Table 7-2: Measured Values And Acceptance

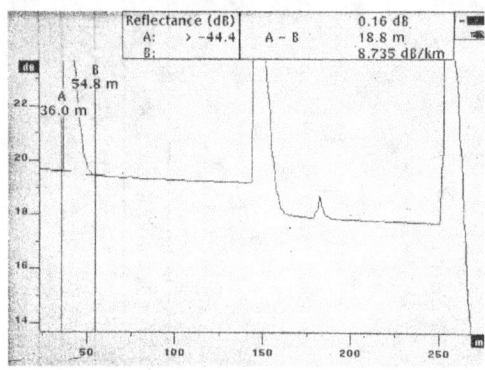

Figure 7-9

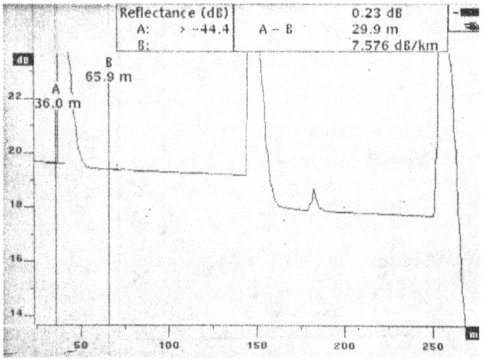

Figure 7-10

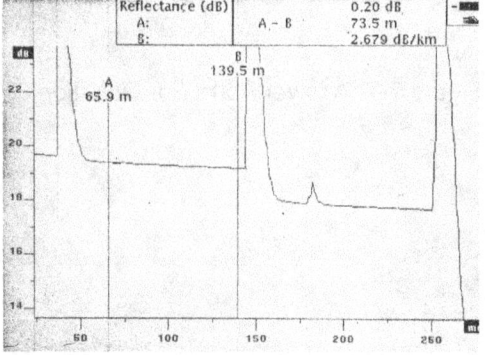

Figure 7-11

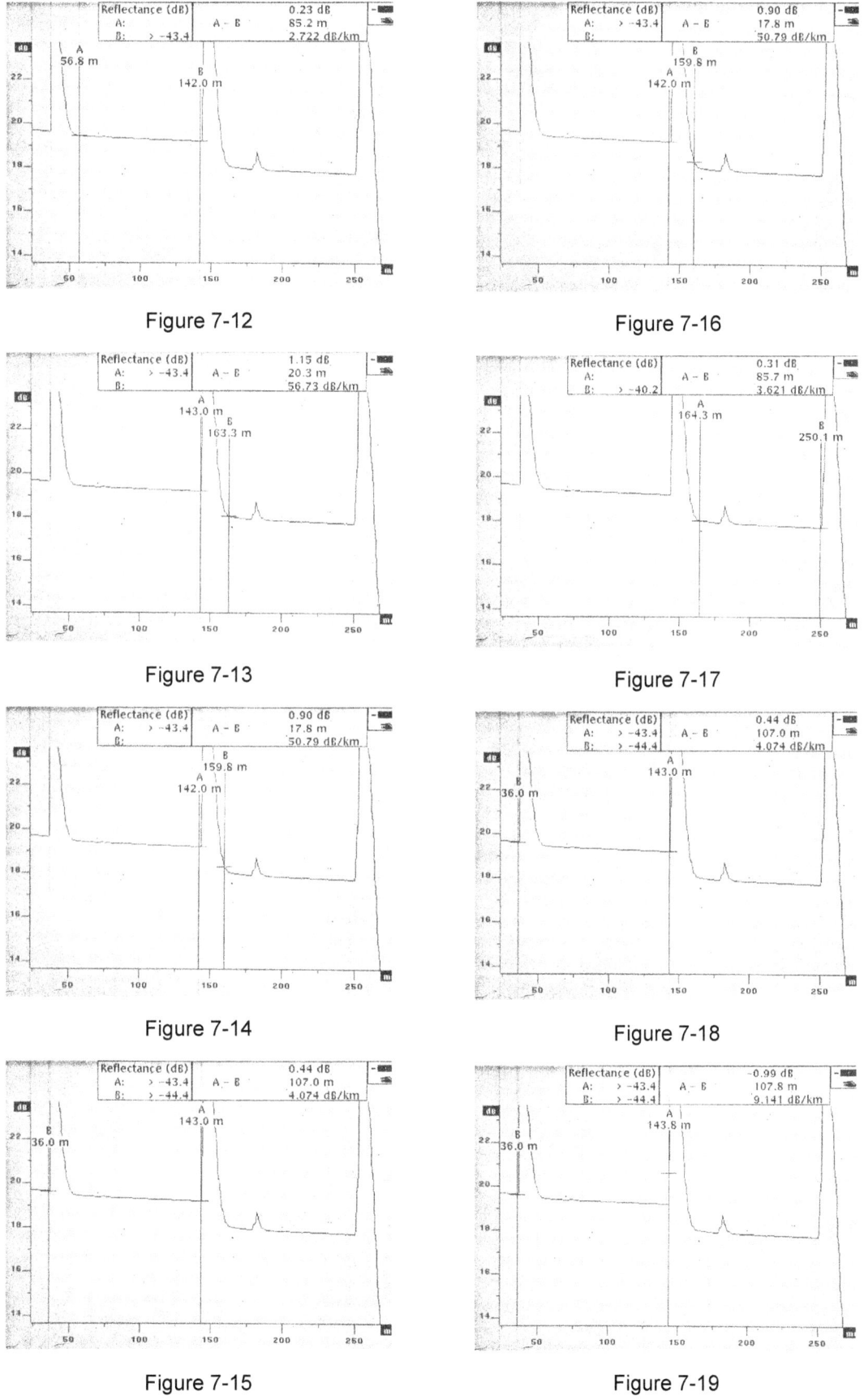

Figure 7-12

Figure 7-16

Figure 7-13

Figure 7-17

Figure 7-14

Figure 7-18

Figure 7-15

Figure 7-19

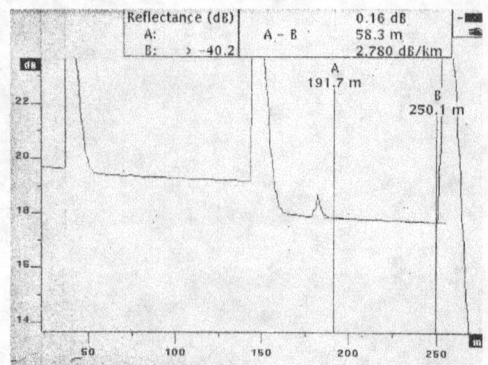

Figure 7-20

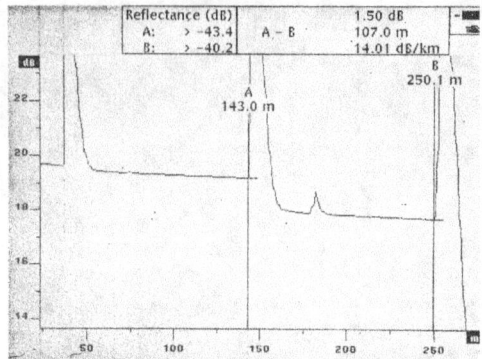

Figure 7-21

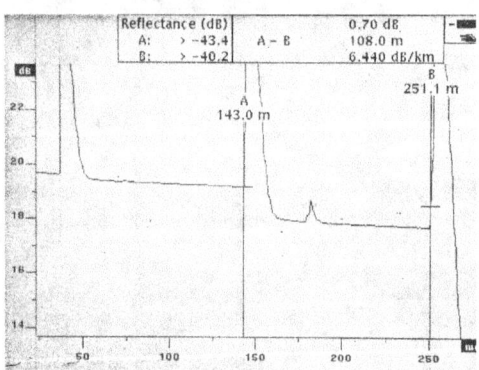

Figure 7-22

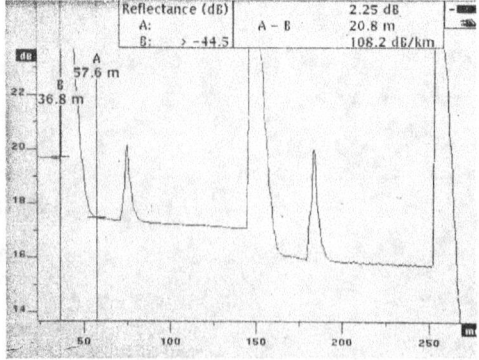

Figure 7-23: End 2

8 APPENDICES

8.1 INDICES OF REFRACTION

8.1.1 MULTIMODE

		Wavelength=	850	1300
Corning	Clearcurve		1.480	1.479
OFS	LaserWave/OM4/OM3		1.483	1.479
Prysmiam/Draka			1.482	1.477

8.1.2 SINGLEMODE

		Wavelength=	1310	1550
Corning	28e+		1.4676	1.4682
	LEAF			1.4680
OFS	Allwave		1.4670	1.4680
	TruWave Reach		1.4710	1.4700
Prysmiam/Draka	G.652		1.4670	1.4680
	Teralight/G.655		1.4682	1.4683

8.2 BACKSCATTER COEFFICIENTS

8.2.1 MULTIMODE

		Wavelength=	850	1300
Corning	Clearcurve		-68.0	-76.0
OFS	LaserWave/OM4/OM3		-68.4	-75.8

8.2.2 SINGLEMODE

		Wavelength=	1310	1550	1625
Corning	28e+		-77.0	-82.0	
	LEAF			-81.0	-82.0
Prysmiam/Draka	G.652		-79.4	-81.7	-82.5
	Teralight/G.655		-77.4	-80.4	-81.3

8.3 IR INACCURACIES

Error in distance to event

Value Entered	Amount of Error		Error, In m @	Error, In m @	Error, In m @	Error, In m @	Error, In m @
1.482	IR error	% error	100 m	500 m	1000 m	5000 m	10000 m
Actual IR			0	0	0	0	0
1.483	0.001	0.0674309	0.07	0.34	0.67	3.37	6.74
1.484	0.002	0.1347709	0.13	0.67	1.35	6.74	13.48
1.485	0.003	0.2020202	0.20	1.01	2.02	10.10	20.20
1.486	0.004	0.2691790	0.27	1.35	2.69	13.46	26.92
1.487	0.005	0.3362475	0.34	1.68	3.36	16.81	33.62
1.488	0.006	0.4032258	0.40	2.02	4.03	20.16	40.32
1.489	0.007	0.4701142	0.47	2.35	4.70	23.51	47.01

8.4 DISTANCE INACCURACIES

Difference between fiber length and cable length, in meters. Cable length is less than fiber length.

fiber excess length	buffer tube excess length	100 m	500 m	1000 m	5000 m	10000 m
0.01%	1.91%	1.92	9.60	19.19	95.97	191.94
0.02%	1.91%	1.93	9.65	19.29	96.47	192.94
0.03%	1.91%	1.94	9.70	19.39	96.97	193.94
0.04%	1.91%	1.95	9.75	19.49	97.47	194.94
0.05%	1.91%	1.96	9.80	19.59	97.97	195.94
0.06%	1.91%	1.97	9.85	19.69	98.47	196.94
0.07%	1.91%	1.98	9.90	19.79	98.97	197.94
0.08%	1.91%	1.99	9.95	19.89	99.47	198.94
0.09%	1.91%	2.00	10.00	19.99	99.97	199.94
0.10%	1.91%	2.01	10.05	20.09	100.47	200.94

8.5 ATTENUATION RATES

Wavelength, nm	Core Diameter, μ	Attenuation Rate, dB/km
850	62.5	3.5
850	50	3.5
1300	62.5	1.5
1300	50	1.5
1310[5]	8.2	0.5
1310[6]	8.2	1.0
15502	8.2	0.5
1550[2]	8.2	1.0

Table 8-1: Maximum Cable Attenuation Rates

Wavelength, nm	Core Diameter, μ	Attenuation Rate, dB/km
850	62.5	2.8-3.0
850	50	2.5-2.7
1300	62.5	0.7
1300	50	0.7
1310	8.2	0.30-0.35
1550	8.2	0.20

Table 8-2: Typical Cable Attenuation Rates

8.6 ACCCEPTANCE VALUES

Wavelength, nm	Core Diameter, μ	Attenuation Rate, dB/km
850	62.5	3.15-3.25
850	50	3.0-3.1
1300	62.5	1.1
1300	50	1.1
1310[7]	8.2	0.4-0.425
1310[8]	8.2	No data
1550[2]	8.2	0.35
1550[2]	8.2	No data

Connector Loss

➢ 0.525 dB/km (polish installation, 2.5 mm ferrule)

➢ 0.475 dB/km (polish installation, 1.25 mm ferrule)

➢ 0.588 dB/km (cleave and crimp installation)

Splice Loss: 0.125 dB/splice

---

[5] Loose tube cable
[6] Tight tube cable

[7] Loose tube cable
[8] Tight tube cable

8.7   BI-DIRECTONAL LENGTH MEASUREMENTS

From	To	1310 nm length, ft		Difference	1550 nm length, ft		Difference
		forward	reverse	ft	forward	reverse	ft
M1	E	43,459	43,476	-17	43,442	43,476	-34
M1	E	43,408	43,476	-68	43,442	43,476	-34
M1	E	142,290	142,324	-34	142,324	142,324	0
M2	T	142,256	142,290	-34	142,290	142,324	-34
M3	SC	80,191	80,191	0	80,242	80,242	0
M3	SC	80,174	80,174	0	80,208	80,208	0
P	M3	104,070	104,070	0	104,104	104,087	17
P	M3	104,070	104,070	0	104,104	104,104	0
M4	ER	157,409	157,510	-101	157,442	157,409	33
SC	L	74,856	75,202	-346	74,890	75,236	-346
SC	L	74,856	75,371	-515	75,422	75,236	186
S1	M4	88,782	88,765	17	88,748	88,816	-68
S1	M4	88,697	88,782	-85	88,748	88,647	101
T	M3	180,746	180,746	0	180,818	180,848	-30
T	M3	180,713	180,713	0	180,814	180,814	0
W	M1	77,536	77,536	0	77,586	77,570	16
W	M1	77,536	77,536	0	77,570	77,553	17
W	M1	77,502	77,536	-34	77,570	77,586	-16
W	M1	77,536	77,536	0	77,586	77,553	33
E	S2	84,334	84,283	51	84,266	84,334	-68
E	S2	84,250	84,334	-84	84,266	84,334	-68
W	M2	107,148	107,148	0	107,181	107,181	0
W	M2	107,181	107,181	0	107,215	107,215	0
M4	A	122,842	122,842	0	122,875	122,875	0
M4	A	122,842	122,808	34	122,842	122,825	17
WD	M4	122,861	122,845	16	122,912	122,912	0
WD	M4	122,861	122,845	16	122,895	122,879	16
CG	M5	68,674	68,708	-34	68,742	68,759	-17
CG	M5	68,708	68,708	0	68,742	68,742	0
CG	M5	68,708	68,708	0	68,742	68,742	0
CG	M5	68,708	68,674	34	68,708	68,742	-34

8.8   BIDIRECTIONAL SPLICE LOSS TESTING

Event	End 1 real distance	loss dB	reverse distance	delta kft	End 2 real distance	loss dB	average loss dB
1	0	0.27	164.227	-0.068	164.295		0.13
2	5.37	0.23	158.857				0.12
3	21.453	0.05	142.774				0.03
4	35.82	-0.06	128.407	-0.017	128.424	0.22	0.08
5					125.247	0.14	0.07
6	42.765	0.18	121.462				0.09
7	43.852	0.33	120.375	0.045	120.33	0.12	0.18
8	46.553	-0.26	117.674	-0.035	117.709	0.24	-0.01
9	51.276	0.31	112.951	0.091	112.86	-0.11	0.10
10	57.169	0.07	107.058	0.191	106.867	-0.06	0.01
11					103.36	0.22	0.11
12	62.875	0.17	101.352	0.262	101.09	0.19	0.18
13	79.508	-0.08	84.719	0.255	84.464	0.06	-0.01
14	90.652	0.26	73.575				0.13
15	97.557	-0.31	66.67	-0.011	66.681	0.32	0.01
16	104.458	0.07	59.769	0.017	59.752	0.10	0.08
17	109.227	0.07	55				0.04
18	115.841	0.15	48.386				0.07
19					45.793	0.08	0.04
20					38.567	0.16	0.08
21					26.866	0.19	0.10
22					14.909	0.07	0.04
23					13.45	-0.14	-0.07
24					10.611	0.13	0.06
25					3.512	0.09	0.04
26	162.686	0.14	1.541				0.07
End	164.227				0		

15

gainers=	5		gainers=	3
ave.=	0.08 dB		ave.=	0.11 dB

AVERAGE
0.07
dB

8.10 REVIEW QUESTIONS ANSWERS

For most questions, we provide the section[s] and figure[s] in which the answer is. Where necessary, we provide the exact answer.

8.10.1 CHAPTER 1

1. Table 1-1
2. 1.1.4
3. 1.3
4. 1.3
5. 1.3
6. 1.2.2
7. 1.1.5
8. 1.1.5
9. 1.1.5
10. 1.1.7
11. 1.1.7
12. 1.1.7, Appendix 8.1
13. 1.1.7
14. 1.1.8
15. 1.1.7, 1.1.9
16. 1.2.1
17. 1.2.1
18. Figure 1-11
19. Figure 1-11
20. Figure 1-11
21. 1.1.7, 1.8.2
22. 1.7
23. 1.7
24. 1.8.1
25. 1.8.1
26. 1.8.1
27. 1.8.2
28. 1.8.3, 1.8.4
29. 1.8.4
30. 1.8.3, 1.8.4
31. 1.8.3
32. 1.1.4

8.10.2 CHAPTER 2

1. 2.2.1
2. 2.1
3. 2.2
4. 2.2.1
5. 2.2.3
6. 2.2.2
7. 2.3.1
8. 2.3.1
9. 2.3.3
10. 2.2.4
11. 2.3.4
12. 2.2.5
13. Figure 2-12, 2.3.1
14. 2.2.1-2.2.5
15. 2.3.1-2.3.3
16. 2.3.2-2.3.4
17. 2.2-2.3
18. 2.2-2.3
19. 2.3.3, 3.2
20. 2.4

8.10.3 CHAPTER 3

1. 3.2
2. 3.2

3. 3.3.4

4. 3.3.4

5. 3.3.2

6. 3.3.2

7. 3.6

8. 3.5

9. 3.6, Table 1-3

10. 3.6

11. 3.4.2

12. 3.4.2

13. 3.4.2

14. 3.4.2

15. 3.4.3

16. 3.4.3

8.10.4   CHAPTER 4

1. 4.1.1

2. 4.1.1

3. 4.1.1.7

4. 4.1.2.1

5. 4.1.2.1

6. 4.1.2.1

7. 4.1.2.2

8. 4.1.2.2

9. 4.1.2.3

10. 4.1.2.3

11. 4.1.2.3

12. 4.1.2.3

13. 4.2.2

14. 4.2.4

15. 4.2.3.1

16. 4.2.3.3

17. 4.2.2

18. 4.2.3.1

19. 4.2.3.1

20. 4.2.5

21. 4.2.5

22. 4.2.3.2

23. 4.2.3.2

24. 4.2.3.3, Figure 4-9

25. 4.2.4, Figure 4-17, Figure 4-21

26. 4.2.3.3, Figure 4-16

27. 4.2.4, Figure 4-21

28. 4.2.5

29. 4.2.3.1, Figure 4-7

30. 4.2.3.1, Figure 4-7

31. 4.2.2, Figure 4-4

32. From end 1

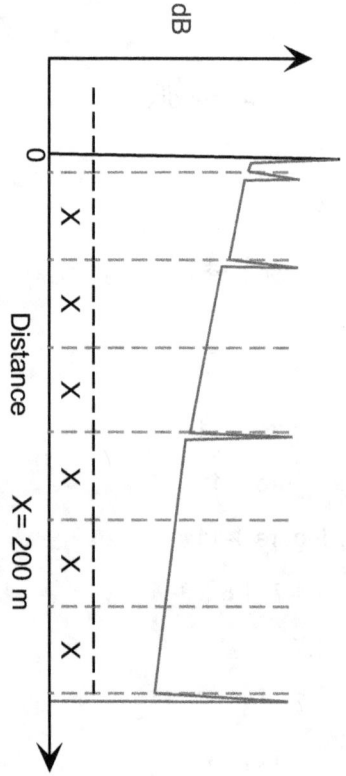

33. From end 2

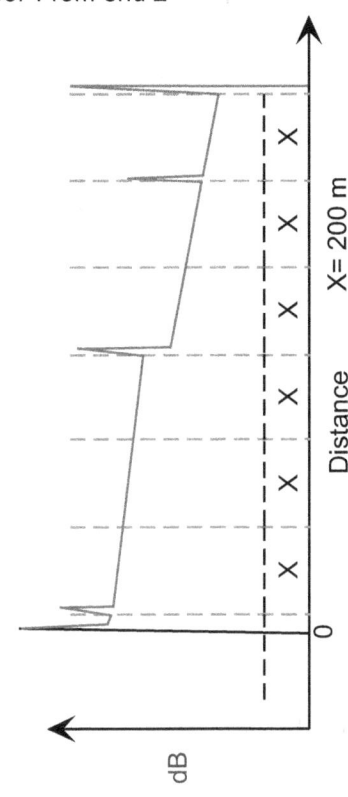

35. From end 2 with ghost

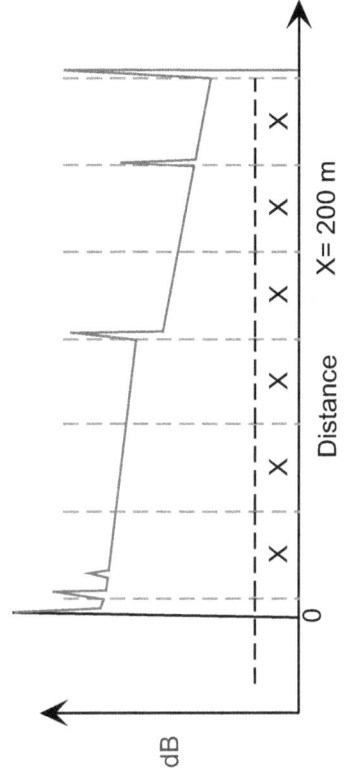

34. From end 1 with ghost

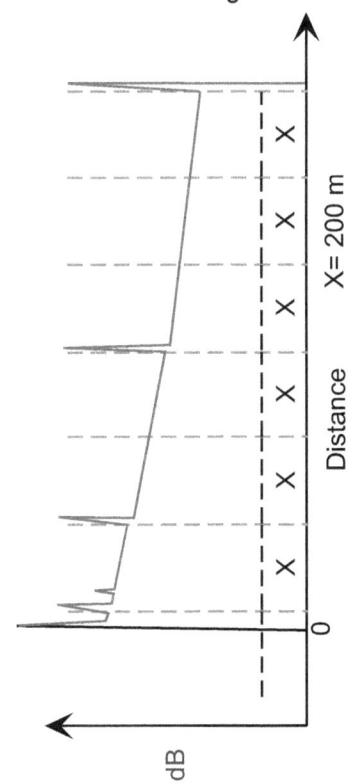

36. From end 1

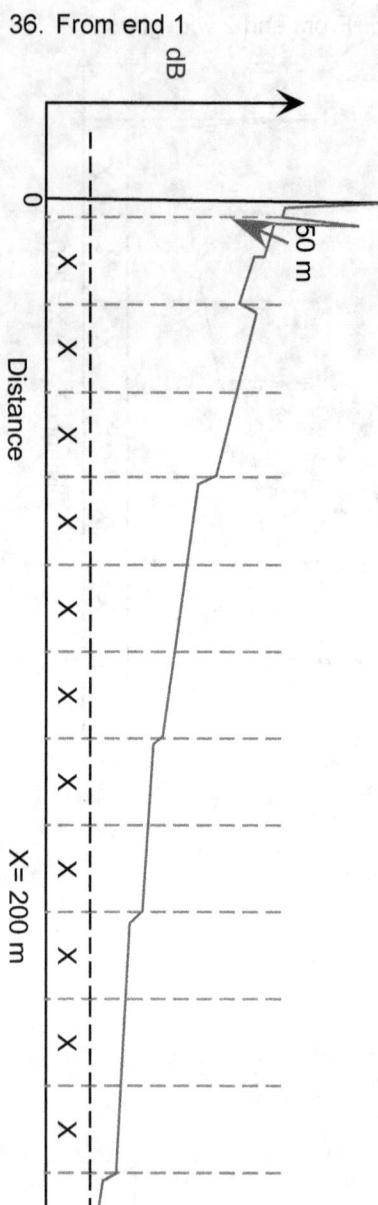

37. From end 2

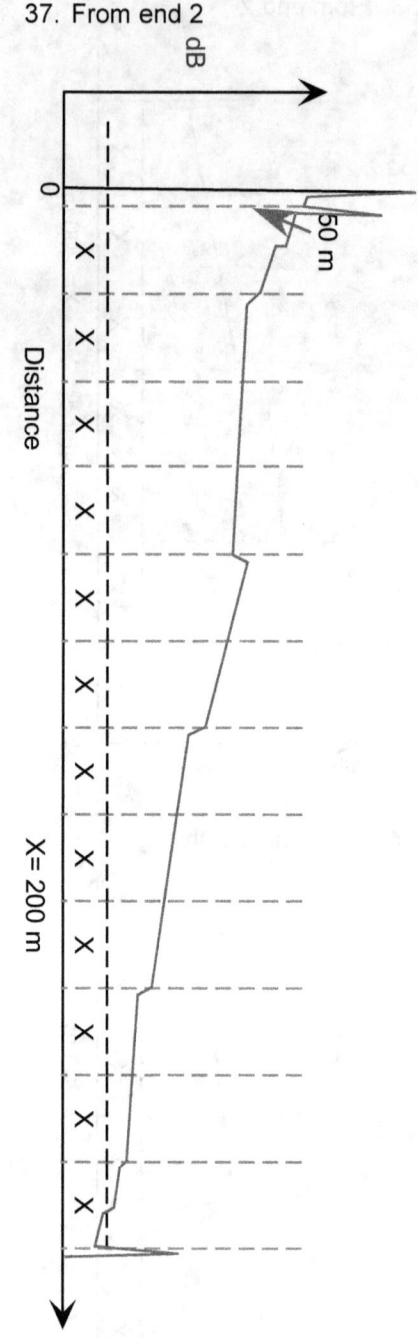

8.10.5   CHAPTER 5

1. Cursor Placement Evaluation
   Results

Figure	Valid?	Type
7-9	Yes	Connector loss
7-10	No	
7-11	Yes	Attenuation rate
7-12	Yes	Attenuation rate
7-13	Yes	Connector loss
7-14	No	
7-15	Yes	Length
7-16	No	
7-17	No	
7-18	Yes	Length
7-19	No	
7-20	Yes	Attenuation rate
7-21	Yes	Length
7-22	No	
7-23	Yes	Connector loss

2. Acceptance Results

	Fig. #	Value	Accept?
Connector End 1	7-9	.16	Yes
Connector End 2	7-23	2.25	No
Connector, Middle	7-13	1.15	No
Length, Segment 1	7-15, 7-18	107.0	Yes
Length, Segment 2	7-21	107.0	Yes
Attenuation Rate, Segment 1	7-11	2.679	Yes
Attenuation Rate, Segment 2	7-20	2.780	Yes

8.11 GLOSSARY

905 SMA: see SMA 905

adapter: a device for mating two connectors. Also known as mating adapter, barrel, bulkhead and feed through

APD: avalanche photodiode. This device converts an optical signal to an electrical signal.

armor: a layer of material, usually stainless steel, which is placed around a cable core to prevent damage from gnawing rodents and from crush loads.

attenuation: the loss of optical power or intensity as light travels through a fiber. It is expressed in units of decibels. When used to describe fibers or cables, it is expressed as a rate in decibels/kilometer. In this book, attenuation refers to reduction in signal strength in the fiber or cable.

back reflection: an outdated term that was used to mean return loss or reflectance. See reflectance and return loss.

back shell: the portion of a connector in back of the retaining nut or latching mechanism.

bandwidth: a measure of the transmission capacity of an analog transmission system.

bandwidth-distance product: the product of the length of a fiber and the analog bandwidth that the fiber can transmit over that length. It is expressed in units of MHz-km for multimode fibers. It is not the relevant parameter for laser optimized fibers. It is not used for singlemode fibers.

bend radius, long term: see bend radius, minimum unloaded.

bend radius, minimum loaded: the smallest radius to which a cable can be bent during installation at the maximum recommended installation load without any damage to either the fiber or the cable materials. Typically, this radius is 20 times the diameter.

bend radius, minimum unloaded: the smallest radius to which a cable can be bent without any damage to either the fiber or the cable materials while the cable is unloaded. Typically, this radius is 10 times the diameter for the life of the cable.

bend radius, short term: see bend radius, minimum loaded.

biconic: a type of connector.

binding tape: a cable component. This tape holds buffer tubes together during jacket extrusion.

binding yarn: a cable component. This yarn holds buffer tubes together during jacket extrusion.

bit error rate: a measure of the accuracy of a digital fiber optic system. The BER is the rate of errors produced by the optoelectronics. Abbreviated as BER.

bit rate: the data transmission rate of a digital transmission system.

boot: a plastic device that slides over the cable and the back shell. It serves to limit the radius of curvature of the cable as the cable exits the back shell.

break out: a type of cable composed of sub cables, each of which contains a single fiber.

3buffer coating: obsolete term. See 4primary coating.

buffer tube: a layer of plastic that surrounds a fiber or a group of fibers.

bulkhead: see adapter.

butt coupling: a method of transmitting light from one fiber to another by precise mechanical alignment of the two fiber ends without the use of lenses.

BWDP: bandwidth-distance product.

cable: the structure that protects an optical fiber or fibers during installation and use.

cable core: the structure of fibers, buffer tubes, fillers and strength members that reside inside the inner most jacket of a cable.

cable end boxes: the enclosures placed on the end of a cable to protect the buffer tubes and fibers.

cap: a plastic structure that protects the end of a connector ferrule from dust and damage when the connector is not in use.

CBT: see central buffer tube

central buffer tube: a cable design in which all fibers reside in a single, centrally located buffer tube.

central strength member: a strength member that resides in the center of a cable.

chromatic dispersion: the spreading of pulses of light due to rays of different wavelengths traveling at different speeds through the core. See spectral width.

cladding diameter: the outer diameter of the cladding. It is measured in micrometers.

cladding non-circularity: the degree to which the cladding and the core deviate from perfect circularity.

cladding: the region of an optical fiber that confines the light to the core and provides additional strength to the fiber.

cleaver: device to create a flat and perpendicular surface on end of a fiber.

cleaving: the process of creating a fiber end that is flat and perpendicular to the axis of the fiber.

coating: see primary coating

concentricity: the degree to which the core deviates from being in the exact center of the cladding.

cone of acceptance: the cone defined by the critical angle or the numerical aperture. This is the cone within which all of the light enters and exits a fiber.

core diameter: the diameter of the region in which most of the light energy travels. It is measured in micrometers.

core offset: see concentricity.

core: the region of an optical fiber in which most of the light energy travels.

coupler: a device that allows two separate optical signals to be joined for transmission on a single fiber.

crimp ring: the device that is deformed around the back shell of a connector. The crimp ring traps the strength member of the cable, providing acceptable cable-connector strength.

critical angle: the maximum angle to the axis of a fiber at which rays of light will enter a fiber and experience total internal reflection at the core- cladding boundary.

crush load, maximum recommended: the recommended maximum load that can be applied to a fiber optic cable without any permanent change in the attenuation of the cable. This can be specified as a long-term or a short-term crush load.

cut-off wavelength: the wavelength, below which a singlemode fiber will not transmit a single mode. Below this wavelength, the singlemode fiber will transmit multimode manner.

D4: a connector type

design: used in this text to refer to a cable

dielectric: having no components that conduct electricity in the cable.

differential modal attenuation (DMA): the mechanism by which rays in multimode fibers experience differing attenuation rates due to their mode, or location in the core.

differential modal delay (DMD): a measurement of dispersion in a multimode fiber; DMD is the measurement used for bandwidth measurement of laser-optimized for fiber.

dispersion: the spreading of pulses in fibers.

ESCON™: a connector type.

expanded beam coupling: a method of transmitting light from one fiber to another with lenses.

FC: a connector type.

FC/PC: a connector type.

FDDI: fiber data distributed data interface.

feed through: see adapter.

ferrule: that portion of the connector that aligns a fiber. A ferrule is present in all connector types except the Volition™.

fiber: the structure that guides light in a fiber optic system.

Fiber Jack: see Opti-Jack™.

filled and blocked cable : a type of cable in which all empty space is filled with compounds to prevent moisture ingress. Filling refers to a gel material in buffer tubes. Blocked refers to grease outside of buffer tubes.

fillers: cable materials that fill otherwise empty space in a cable.

Fresnel reflection: the reflection that occurs when light travels between two media in which the speed of light differs.

FRP strength member: a fiberglass reinforced plastic or epoxy rod that is used as a dielectric strength member in cables. The term 'plastic' may refer to either a polymer plastic or an epoxy.

fusion splice: a splice made by melting two fibers together.

gel-filling compound: a compound placed inside a loose buffer tube to prevent water from contacting the fiber(s) in that buffer tube.

graded index: a type of multimode fiber, in which the chemical composition of the core is not uniform.

HCS™: a hard clad silica fiber.

heat shrink tubing: tubing placed on the back shell of a connector.

HiPPI: high speed, parallel processor interface.

index of refraction (IR, h): the ratio of the speed of light in the material to the speed of light in a vacuum. This is a dimensionless number.

inner duct: a corrugated plastic pipe in which fiber optic cables are placed.

inner jacket: any layer of jacketing plastic that is not the outer-most layer.

installation strength, maximum recommended: the maximum load that can be applied along the axis of a cable without any damage to the fibers .

installation temperature range: the temperature range within which a cable can be installed without damage.

jacket: a layer of plastic in a cable. It can be an outer jacket or an inner jacket.

jumper: a short length of a single fiber cable with connectors on both ends.

Kevlar®: an aramid yard produced by DuPont Chemical. It is used to provide strength in fiber optic cables.

keying: a connector mechanism by which ferrules are prevented from rotating.

laser diode: a semiconductor that converts electrical signals to optical signals.

laser optimized (LO) a multimode fiber that is designed to enable long transmission distance at 1 Gbps-100 Gbps.

latching mechanism: device that retains a connector to a receptacle or an adapter .

LC: a SFF connector with a 1.25 mm ferrule.

LX.5: a SFF connector with a 1.25 mm ferrule and a built-in dust cover.

LED: light emitting diode. A semiconductor that converts electrical signals to optical signals.

lensed coupling: see expanded beam coupling.

loose buffer tube: a buffer tube with space between the outer diameter of the primary coating and the inner diameter of the buffer tube.

loose tube: a cable design in which the fiber floats loosely inside an oversized tube.

loss: the end to end reduction in optical power as light travels through a fiber, connectors or splices. In this book, the 'loss' refers to reduction of optical power at a splice or a connector pair.

material dispersion: the spreading of pulses of light due to rays of light traveling through different regions of the core.

maximum recommended installation load: see installation strength, maximum recommended.

mechanical splice: a mechanism that aligns two fiber ends precisely for efficient transfer of light from one fiber to another.

MFPT: a multiple fiber per (buffer) tube cable design. This design usually has 6 or 12 fibers per loose buffer tube.

mini-BNC: a connector type.

minimum recommended long term bend radius: see bend radius, minimum unloaded.

minimum recommended short-term bend radius: see bend radius, minimum loaded.

MPO: the generic term for a 4-24 fiber, connector with a single ferrule. The usual form of use is MPO/MPT.

MPT: the proprietary term for a 4-12 fiber, connector with a single ferrule. The usual form of use is MPO/MPT.

modal dispersion: the spreading of pulses of light due to rays traveling different paths through a multimode fiber.

modal pulse spreading: see modal dispersion.

mode: a path in which light can travel in a fiber core. Mode translates roughly to 'path.'

mode field diameter: the diameter within which the light energy field travels in a singlemode fiber.

monomode: see singlemode. This term is used outside of North America.

MT-RJ: a duplex SFF connector with a single ferrule.

MU: a SFF connector with a 1.25 mm ferrule and a size of approximately half that of the SC connector.

multimode: a method of propagation of light in which all of the rays of light do not travel in a path parallel to the axis of the fiber.

NA: see numerical aperture.

numerical aperture (NA): the sine of the critical angle. The NA is a measure of the solid angle within which rays of light will enter and exit the fiber.

offset, core: see core offset.

optical amplifier: a device that increases the signal strength without an optical to electrical to optical conversion.

optical coupling: see expanded beam coupling.

optical power budget: the maximum loss of optical power which transmitter-receiver pair can withstand while still functioning at the specified level of accuracy.

Opti-Jack™: a duplex SFF connector.

optical return loss: see return loss.

optical rotary joint: a rotating joint that allows transmission of light from a stationary fiber to a rotating fiber.

optical switch: a switch that can direct light to more than one output path.

optical time domain reflectometer (OTDR): a test device that creates a map of the loss of signal strength of an optical path.

optical waveguide: another term for an optical fiber.

optoelectronic device: any device that converts a signal from electrical to optical domain or vice versa.

optoelectronics: see optoelectronic device.

OTDR: optical time domain reflectometer or optical time domain reflectometry.

ovality: a measure of the degree to which a fiber deviates from perfect circularity. See cladding non-circularity.

passive component: a device that manipulates light without requiring an optical to electrical signal conversion.

patch panel: a sheet of material that contains adapter(s).

PCS: a plastic clad silica fiber.

PD: photodiode. This device converts an optical signal to an electrical signal.

pigtail: a length of fiber or cable that is permanently attached to a connector or an optoelectronic device.

Ping-Pong: a type of transmission that results from using a single LED as both a transmitter and receiver.

plug: another term for connector.

POF: plastic optical fiber. A fiber with a plastic core and a plastic cladding.

polishing fixture: device to hold connector perpendicular to fiber end;

polishing puck: see polishing fixture

polishing tool: see polishing fixture

primary coating: a layer of plastic placed around the cladding by the fiber manufacturer.

primary coating diameter: the diameter of the layer of plastic that is placed around the fiber by the fiber manufacturer.

pull-proof: a performance characteristic of connector types. A connector type is pull-proof when tension on the cable attached to the connector does not produce an increase in the loss of the connector.

pulse dispersion: see dispersion.

pulse spreading: see dispersion.

receptacle: the device within which an active device is mounted. The receptacle is designed to mate with a specific connector type.

reflectance: a measure of the ratio of reflected power to incident power for a single device, such as a connector or mechanical splice. Reflectance is measured in units of dB.

reinforced jacket: two layers of plastic that are separated by strength members.

repeatability: the maximum change in loss between successive measurements of the loss of a connector pair.

retaining nut: device that attaches a connector to receptacle or to an adapter

return loss: the ratio of incident power to power reflected or back scattered for an entire.

ribbon: a structure on which multiple fibers are precisely aligned.

SAP: see super absorbent polymer

super absorbent polymer: a material that absorbs moisture by converting it to a gel; used in fiber optic cables to provide moisture resistance; incorporated into cables as tapes, yarns, threads and powders.

SC: a connector type.

SFF: see 'small form factor'.

shrink tubing: plastic that covers back shell of connector

simplex: a single fiber cable.

singlemode: a method of propagation of light in which all of the energy of light travels the same path length

slotted core: a cable design with a core containing helical slots.

SMA 905: a connector type

SMA 906: a connector type

small form factor: a type of connector with the characteristic of small size that enables doubling of density in patch panels, enclosures, and switches.

spectral width: the measure of the width of the output power-wavelength curve at a power level equal to half the peak power.

splice enclosure: a structure that encloses and protects splice trays and cable ends.

splice tray: a structure that encloses and protects fibers.

splice: a device for permanent alignment of two fiber ends.

splitter: a device that creates multiple optical signals from a single optical signal.

spot size: the size of the area of an LED or laser diode from which light is produced.

ST®: the first of a series of connector types designed by ATT. Other types from ATT are the ST-II and the ST-II+.

star core: see slotted core.

ST-compatible: a connector with a type which is compatible with the ST® connector.

step index: a type of multimode fiber, in which the chemical composition of the core is uniform.

storage temperature range: the temperature range within which a cable can be stored without damage.

strength members: those elements of a cable design that provide strength.

type: the sum of characteristics that differentiates one connector type from another type.

TECS™: a technically enhanced clad fiber similar to hard clad silica fiber.

temperature operating range: the range of temperature within which the cable can be operated during its lifetime without degradation of its properties.

tight tube: a design in which the tube does contact the entire circumference of the fiber. A tight tube can contain only 1 fiber.

total internal reflection: the mechanism by which optical fibers confine light to the core.

type: used in this text to refer to a fiber or connector

use load, maximum recommended: the maximum longitudinal load that can be applied to a cable during its entire lifetime without damage.

VCSEL: vertical cavity, surface emitting laser. A type of relatively low cost multimode light source that transmits at 1-10 Gbps.

Volition™: a duplex SFF connector with no ferrules.

Water-blocking compound: a compound placed in the interstices between buffer tubes or between jackets in a cable.

wavelength division multiplexer: a passive device that combines optical signals with different wavelengths on different fibers light onto a single fiber. The wavelength division demultiplexer performs the reverse function.

wavelength: a measure of the color of the light. It is a length stated in nanometers (nm) or in micrometers (µ).

wiggle proof: a connector performance characteristic. A connector type is wiggle proof if lateral pressure on the back shell does not produce an increase in the loss.

window: the wavelength range within which a fiber is designed to provide specified performance.

Zip cord®: a two-fiber cable with a figure 8 cross-section that allows each of the two fibers to be separated in the same manner as a lamp cord.

8.12 ACRONYMS

ADSS	all dielectric self support
APC	angled physical contact connector
ATM	Asynchronous Transfer Mode
BER	bit error rate
CATV	Cable TV
CWDM	coarse wavelength division multiplexing
D4	a connector
DAC	dual attachment concentrator
DAS	dual attachment station
DS	dispersion shifted
DS-NZD	dispersion shifted, non-zero dispersion
DWDM	dense wavelength division Multiplexing
EDFA	erbium doped fiber amplifier
ESCON	Enterprise System Connection
FC	a connector type
FDDI	fiber data distributed interface
FOLS	Fiber Optic LAN Section of the TIA
FTTD	fiber to the desk
FTTH	fiber to the home
FTTP	fiber to the premises
FTTX	any of the above three
GI	graded index
HDPE	high-density polyethylene

IP	Internet Protocol
LAN	local area network
LD	laser diode
LEAF™ (Corning Inc.)	large effective area fiber
LED	light emitting diode
LSA	least squares analysis
LX.5	a connector type
MFPT	multiple fiber per tube
MIC	media interface connector
MM	multimode
MT-RJ	a duplex connector type
MU	a SFF connector type
NA	numerical aperture
NDS	non-dispersion shifted
NEC	National Electrical Code
NIST	National Institute of Science and Technology
OFC	optical fiber cable, conductive, horizontal rated
OFCP	optical fiber cable, conductive, plenum rated
OFCR	optical fiber cable, conductive, riser rated
OFN	optical fiber cable, non-conductive, horizontal rated
OFNP	optical fiber cable, non-conductive, plenum rated
OFNR	optical fiber cable, non--conductive, riser rated
OPBA	optical power budget available
OPBR	optical power budget requirement
PC	physical contact
PMD	polarization mode dispersion
PON	passive optical network
SAN	storage area network
SAP	super absorbent polymer
SAC	single attachment concentrator
SAS	single attachment station
SC	a connector type
SDH	Synchronous Digital Hierarchy
SI	step index
SM	singlemode
SONET	synchronous optical network
STP	shield twisted pair
UPC	ultra physical contact
UTP	unshielded twisted pair
VCSEL	vertical cavity, surface emitting laser
WDM	wavelength division multiplexing
ZDW	zero dispersion wavelength

8.13 THE AUTHOR

For the last 34 years Mr. Eric R. Pearson has been worked in fiber optic communications. This involvement includes a wide variety of activities, as detailed below.

Mr. Pearson has developed and run two fiber optic cable manufacturing facilities and organizations (Manufacturing Manager, Times Fiber Communications and Business Manager, Whitmor Waveguides). In these positions, Mr. Pearson developed cable designs, manufacturing techniques and qualified designs against performance specifications. As business manager, he was responsible for developing a profitable multi-million dollar business unit.

Mr. Pearson managed two fiber-manufacturing facilities, one for Corning Glass Works and for Times Fiber Communications.

Mr. Pearson has delivered 507 training presentations and trained more than 8400 personnel in proper installation and design procedures. Between his field installations and training, he has made and supervised more than 46,648 connectors. Mr. Pearson has performed tens of thousands of OTDR, ORL, insertion loss, and dispersion tests on both multimode and singlemode cables.

From both field experience and training, Mr. Pearson gained sufficient experience to write three definitive texts on cable and connector installation, The Complete Guide to Fiber Optic Cable Installation (Delmar Publishers, 1997, ISBN #0-8273-7318-X), Successful Fiber Optic Installation- The Essentials (2005), and this text.

He has written the books: Fiber Optic Network Design, Practical Fiber Optic System Design and Implementation, How to Specify and Choose Fiber Optic Cables, and How to Specify and Choose Fiber Optic Connectors.

Mr. Pearson has been a technical expert for a patent infringement suits, law suits between operators and end users, law suits regarding technical fraud in fiber optic cables, and performance specifications.

From 1995-1997, Mr. Pearson was a Director and the Director of Certification, of the Fiber Optic Association (FOA). As the latter, he is responsible for developing requirements and examinations for basic to advanced certification of fiber optic operator personnel. These activities require an in-depth knowledge of all aspects of cable and connector installation.

From the Fiber Optic Association, he has received four advanced Certified Fiber Optic Specialist certifications (CFOS/T, CFOS/S, CFOS/C, and CFOS/I). In 2011, FOA has designated Mr. Pearson as a 'Master Instructor'.

Since 1999, Mr. Pearson has been a Master Instructor for the Building Industry Consult-ants Services International (BICSI), and the developer of the BICSI fiber optic network design program.

From 1986-2004, he was been a Member, Editorial Advisory Board, Fiberoptic Product News.

The Academy of Professional Consultants & Advisors (APCA) certified him as a Certified Professional Consultant (CPC).

He has over 100 articles, reports and presentations to his credit. He is frequently quoted in fiber optic and related trade journals.

He has been selected to speak at three Newport Fiber Optic Marketing Conferences.

He is listed in: Who's Who Worldwide, Who's Who of Business Leaders, Who's Who in Technology and Who's Who in California.

Mr. Pearson has provided consulting services to hundreds of companies in the areas of fiber network design and specification, technical and marketing evaluations.

Mr. Pearson received his education at Massachusetts Institute of Technology (BS, 1969) and Case-Western Reserve University (MS, 1971). Both degrees are in Metallurgy and Materials Science.

Mr. Eric R. Pearson, CFOS

8.14 SERVICES

Training Programs
(http://www.ptnowire.com/training-list.htm)

Basic Installation (4 days)
http://www.ptnowire.com/Fpro1.htm

Basic Installation With FOA CFOT
Certification (5 days)

Basic With Advanced Connector
Installation (5 days)
http://www.ptnowire.com/Fpro1-2.htm

Advanced Connector Installation (2 days)
http://www.ptnowire.com/Fpro2.htm

Advanced Testing (3 days)

Advanced Splicing (5 days)

Advanced With FOA CFOS Certification

Sales Training

Books

Professional Fiber Optic Installation,
The Essentials For Success

Mastering The OTDR- Trace
Acquisition And Interpretation

Mastering Fiber Optic Testing (in
development)

Fiber Optic Network Design (in
development)

Consulting
(http://www.ptnowire.com/services.htm)

Network Design Review

Lawsuit Technical Support

Technical Support For Marketing

Used Splicing And Test Equipment

INDEX

www.ingramcontent.com/pod-product-compliance
Lightning Source LLC
Chambersburg PA
CBHW081213170526
45165CB00009B/2802